Advances in Computer Vision and Pattern Recognition

More information about this series at http://www.springer.com/series/4205

Jean-Denis Durou · Maurizio Falcone ·
Yvain Quéau · Silvia Tozza
Editors

Advances in Photometric 3D-Reconstruction

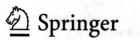 Springer

Editors
Jean-Denis Durou
IRIT, UMR CNRS 5505
University of Toulouse
Toulouse, France

Maurizio Falcone
Department of Mathematics
Sapienza University of Rome
Rome, Italy

Yvain Quéau
GREYC Laboratory
CNRS
Caen, France

Silvia Tozza
Department of Mathematics and
Applications "Renato Caccioppoli"
University of Naples Federico II
Naples, Italy

ISSN 2191-6586   ISSN 2191-6594   (electronic)
Advances in Computer Vision and Pattern Recognition
ISBN 978-3-030-51868-4   ISBN 978-3-030-51866-0   (eBook)
https://doi.org/10.1007/978-3-030-51866-0

This Springer imprint is published by the registered company Springer Nature Switzerland AG
The registered company address is: Gewerbestrasse 11, 6330 Cham, Switzerland

Preface

The Shape-from-X class of problems is related to the 3D-reconstruction of the shape of objects, an issue largely investigated by the computer vision and applied mathematics communities. The X represents a bunch of information that can be used to reconstruct the shape, e.g. from shading, from texture, and from polarization.

To this class belongs the classic Shape-from-Shading (SfS) problem, where the datum is a single 2D gray-level image of the object we want to reconstruct. This approach dates back to the pioneering work of Horn in the seventies. Usually, all these 3D-reconstruction problems can be formulated via partial differential equations and/or via variational methods, giving rise to a variety of nonlinear systems that have been analyzed by many authors. A typical feature is that all the methods require advanced techniques for the analysis and for the numerical approximation since 3D-reconstruction is a typically ill-posed problem that, in general, does not admit a unique solution. Moreover, for real applications we need to reconstruct objects of different materials and shapes having different reflection properties and corresponding to non-regular surfaces. To this end, various models have been proposed and some of them are assuming known position of light sources (calibrated model), whereas others are not (uncalibrated models). Clearly, real-world applications need the analysis of uncalibrated models.

The ingredients that play a role in this context are essentially three: the camera model, the reflectance model, and the lighting model. The most simple and common setup found in the literature is the Lambertian model under orthographic projection, but perspective models and non-Lambertian surfaces have also been studied. In order to solve the under-determination, one can couple the basic model with other information as, for example, using more images of the same object. For instance, in the photometric stereo technique, one uses several images taken from the same point of view under different lighting conditions, whereas with a stereovision technique the images are obtained while moving the viewpoint (camera) under the same light configuration.

This volume is devoted to photometric 3D-reconstruction techniques (shape-from-shading, photometric stereo, and shape-from-polarization), and our goal is to present recent advances in the area. Some contributions will focus on theoretical results which are necessary to develop real-world applications, a field that is growing in recent years, thanks to the advances on effective numerical techniques. However, the features listed above imply serious difficulties at a numerical level, for both accuracy and complexity, and further research activity is needed to make a step forward in their application.

Our aim is to give an overview of photometric techniques that can be useful by researchers, engineers, practitioners, and graduate students involved in computer vision. Also researchers in optimization and numerical analysis can find this volume interesting to approach the field.

The papers presented in this volume give only an idea of the different research directions discussed during the Workshop "Recent Trends in Photometric 3D-Reconstruction" held in Bologna within the SIAM Conference on Imaging Science in June 2018. A detailed description of the papers is presented at the end of the first chapter.

We take this opportunity to thank all the speakers of that workshop and those who contributed to this volume. We hope that it will attract more researchers in this challenging area.

Toulouse, France   Jean-Denis Durou
Rome, Italy   Maurizio Falcone
Caen, France   Yvain Quéau
Naples, Italy   Silvia Tozza
April 2020

Contents

Chapter 1
A Comprehensive Introduction to Photometric 3D-Reconstruction

Jean-Denis Durou, Maurizio Falcone, Yvain Quéau, and Silvia Tozza

Abstract Photometric 3D-reconstruction techniques aim at inferring the geometry of a scene from one or several images, by inverting a physical model describing the image formation. This chapter presents an introductory overview of the main photometric 3D-reconstruction techniques which are shape-from-shading, photometric stereo and shape-from-polarisation.

1.1 Introduction

Inferring the 3D-shape of a scene is necessary in many applications such as quality control, augmented reality or medical diagnosis. Depending upon the requirements of the application, 3D-estimation can be carried out using a variety of technological solutions, from coordinate measuring machines to X-ray scanners. Over the last decades, digital cameras have become a reliable alternative to such sensors, as they represent a reasonable compromise between resolution and affordability. Given one or several 2D-images of a scene captured by a digital camera, the process of estimating its 3D-shape is called *3D-reconstruction*. It is a classic inverse problem in computer vision, which has been addressed in several ways.

J.-D. Durou (✉)
IRIT, UMR CNRS 5505, Université de Toulouse, Toulouse, France
e-mail: durou@irit.fr

M. Falcone
Department of Mathematics, Sapienza University of Rome, Rome, Italy
e-mail: falcone@mat.uniroma1.it

Y. Quéau
CNRS, GREYC Laboratory, ENSICAEN and University of Caen, Caen, France
e-mail: yvain.queau@ensicaen.fr

S. Tozza
Department of Mathematics and Applications "Renato Caccioppoli", Universitá degli Studi di Napoli Federico II, Naples, Italy
e-mail: silvia.tozza@unina.it

© Springer Nature Switzerland AG 2020
J.-D. Durou et al. (eds.), *Advances in Photometric 3D-Reconstruction*,
Advances in Computer Vision and Pattern Recognition,
https://doi.org/10.1007/978-3-030-51866-0_1

1

Table 1.1 Main shape-from-X techniques. Geometric techniques aim at identifying and analysing features. This presentation rather focuses on photometric techniques, which aim at inverting a physics-based image formation model

	Geometric techniques	Photometric techniques
Single image	Structured light [39]	Shape-from-shading (SfS) [48]
	Shape-from-shadows [105]	
	Shape-from-contours [15]	
	Shape-from-texture [125]	
	Shape-from-template [10]	
Multi-images	Structure-from-motion [78]	Photometric stereo (PS) [129]
	Stereopsis [43]	Shape-from-polarisation (SfP) [96]
	Shape-from-silhouettes [44]	
	Shape-from-focus [81]	

A 3D-model consists of a set of *geometric* (position, orientation, etc.) and *photometric* (color, texture, etc.) information. Knowing both these pieces of information allows to render synthetic images, by simulating the trajectory of the light rays from the sources to the camera, after reflection on the surface of the scene. 3D-scanning is the dual of rendering: one aims at a geometric and photometric characterisation of the scene's surface by reversing the trajectory of the light rays. In fact, 3D-scanning includes both the subproblems of 3D-reconstruction (estimating the scene's geometry) and appearance estimation (estimating its photometric properties).

The various 3D-reconstruction techniques from digital cameras are grouped under the generic terminology *shape-from-X*, X indicating that shape estimation can be based on various clues (shadows, contours, etc.). The main shape-from-X techniques are presented in Table 1.1. In this table, they are classified according to the clue they are based on (photometric or geometric) and the number of images they require.

Geometric shape-from-X techniques are built upon the identification of features in the image(s). On the other hand, photometric shape-from-X techniques are based on the analysis of the quantity of light received in each photosite of the camera's sensor. Photometric 3D-reconstruction techniques indeed rely on a physics-based forward image formation model describing the interactions between light, matter and the camera, and aim at inverting this model in order to infer the geometry of the scene and, possibly, its photometric properties.

There exist out-of-the box solutions for geometric 3D-reconstruction, e.g. Microsoft Kinect (based on stereopsis or structured light, depending on the version), or the CMPMVS [54] or AliceVision [2] projects (based on structure-from-motion and stereopsis). On the contrary, there is a lack of such solutions for photometric techniques, which are usually rather viewed as "lab" reconstruction techniques because they rely on several assumptions on the acquisition setup. Still, they bear great promises in terms of level of geometric details which can be recovered, and of applicability to a wide range of materials.

The aim of this chapter is to present an overview of the three main photometric shape-from-X techniques: shape-from-shading, photometric stereo and shape-from-polarisation. We first review in Sect. 1.2 the shape-from-shading problem, which is a computer vision technique consisting in inferring geometry from a single image. Then, we discuss two techniques where multiple images are analysed under controlled incident or reflected lighting. In photometric stereo (Sect. 1.3), a series of images are acquired under varying incident lighting, which permits to estimate both the shape and the reflectance of the pictured surface. In shape-from-polarisation (Sect. 1.4), it is the state of polarisation of the reflected light which is analysed, by considering a series of images acquired with a controllable polarising filter attached to the camera. Section 1.5 eventually concludes this study by presenting the subsequent chapters of this volume.

1.2   Shape-from-Shading

Inferring 3D-geometry from a single image of a shaded surface is a problem known as *shape-from-shading*. This technique was first developed in the seventies at MIT, under the impulse of Horn [48].

1.2.1   Non-differential SfS Models

Let us briefly outline the problem, attaching to the camera a 3D-coordinate system $Oxyz$, such that Oxy coincides with the image plane and Oz with the optical axis. Assuming orthographic projection, the visible part of the scene is, up to a scale factor, a graph $z = u(\mathbf{x})$, where $\mathbf{x} = [x, y]^\top$ is an image point. The SfS problem can be modelled by the *image irradiance equation* [49]:

$$I(\mathbf{x}) = R(\mathbf{n}(\mathbf{x})), \tag{1.1}$$

where $I(\mathbf{x})$ is the graylevel at point $\mathbf{x}$ (in fact, $I(\mathbf{x})$ is the *irradiance* at point $\mathbf{x}$, but both quantities are proportional), and the *radiance* $R(\mathbf{n}(\mathbf{x}))$ gives the value of the light re-emitted by the surface as a function of its orientation, i.e. of the unit normal $\mathbf{n}(\mathbf{x})$ to the surface at the 3D-point $[x, y, u(\mathbf{x})]^\top$ conjugate with $\mathbf{x}$ (cf. Fig. 1.1). Assuming that, besides I, the radiance function R is also known, then solving Eq. (1.1) is a non-differential model of SfS, in which the unknown is the normal $\mathbf{n}(\mathbf{x})$.

Let us assume there is a unique light source at infinity, whose direction is characterised by the unit vector $\boldsymbol{\omega} = [\omega_1, \omega_2, \omega_3]^\top \in \mathbb{R}^3$, and whose intensity is denoted by $\psi(\mathbf{x})$. Let us also assume for simplicity that the surface is *Lambertian*, i.e. an ideal diffuse reflecting surface for which the apparent brightness is independent from the viewing angle. Then, R is written in such a way that Eq. (1.1) becomes

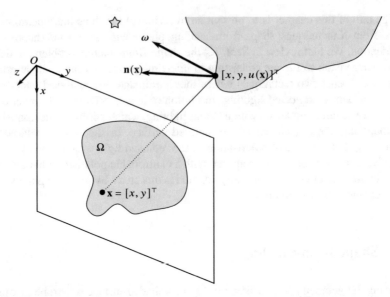

Fig. 1.1 The surface is represented as a graph $z = u(\mathbf{x})$, where $\mathbf{x} = [x, y]^\top$ is an image point in the reconstruction domain Ω. The normal at the surface point $[x, y, u(\mathbf{x})]^\top$ conjugate with $\mathbf{x}$ is denoted by $\mathbf{n}(\mathbf{x})$, and the incident light direction by $\boldsymbol{\omega}$

$$I(\mathbf{x}) = r(\mathbf{x})\,\psi(\mathbf{x})\,\boldsymbol{\omega}^\top \mathbf{n}(\mathbf{x}). \tag{1.2}$$

In Eq. (1.2), $r(\mathbf{x})$ is the *reflectance* (or *albedo*), and the scalar product $\psi(\mathbf{x})\,\boldsymbol{\omega}^\top\mathbf{n}(\mathbf{x})$ is called *shading*. This is another example of non-differential SfS model.

Equation (1.2) is fundamentally ill-posed, according to the *trompe-l'œil* principle, which is well illustrated by Adelson and Pentland's "workshop metaphor" [1] (cf. Fig. 1.2). If a painter, a light designer and a sculptor are asked to design an artwork explaining a given image $I(\mathbf{x})$, they may propose very different, but plausible, solutions. The painter will assume a planar surface and a uniform lighting, the changes in intensity being explained by changes in reflectance $r(\mathbf{x})$. The light designer may propose a sophisticated lighting configuration $\psi(\mathbf{x})$ placed in front of a planar surface with uniform reflectance. Eventually, the sculptor will assume lighting and reflectance are uniform and explain the changes in intensity solely by the shading, which results from variations in the local orientation $\mathbf{n}(\mathbf{x})$ of the surface.

This last explanation, which comes down to inverting Eq. (1.2) in order to infer a 3D-shape, assuming everything is known but $\mathbf{n}(\mathbf{x})$, is precisely the shape-from-shading problem. So it is assumed that the reflectance is known, which is usually written $r(\mathbf{x}) \equiv 1$, and that the lighting is uniform, i.e. $\psi(\mathbf{x}) \equiv 1$.

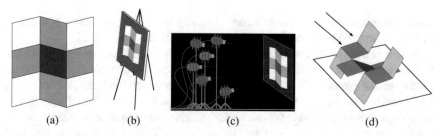

Fig. 1.2 Adelson and Pentland's "workshop metaphor" [1]. To explain an image **a** in terms of reflectance, lighting and shape, **b** a painter, **c** a light designer and **d** a sculptor will design three different, but plausible, solutions. Inferring the shape **d** from a single image is the shape-from-shading problem

1.2.2 Differential SfS Models

Let us now turn to differential SfS models. Under orthographic projection, the normal is easily expressed as

$$\mathbf{n}(\mathbf{x}) = \frac{1}{\sqrt{1 + p(\mathbf{x})^2 + q(\mathbf{x})^2}} [-p(\mathbf{x}), -q(\mathbf{x}), 1]^\top, \qquad (1.3)$$

where

$$p := \frac{\partial u}{\partial x} \quad \text{and} \quad q := \frac{\partial u}{\partial y}, \qquad (1.4)$$

so that $\nabla u(\mathbf{x}) = [p(\mathbf{x}), q(\mathbf{x})]^\top$. It is easily deduced from Eqs. (1.2) and (1.3), assuming $r(\mathbf{x}) \equiv 1$ and $\psi(\mathbf{x}) \equiv 1$, that the following equation holds true for a general parallel lighting whose direction is characterised by $\boldsymbol{\omega} = [\omega_1, \omega_2, \omega_3]^\top$:

$$I(\mathbf{x})\sqrt{1 + |\nabla u(\mathbf{x})|^2} + [\omega_1, \omega_2] \nabla u(\mathbf{x}) - \omega_3 = 0. \qquad (1.5)$$

This is a first-order nonlinear partial differential equation (PDE) of Hamilton–Jacobi type, which constitutes an example of differential SfS model, in which the unknown is now the function u, called the *height map*. This equation has to be solved on a compact domain $\Omega \subset \mathbb{R}^2$, called the *reconstruction domain*.

The PDE which appears in most of the papers on SfS corresponds to a frontal lighting, i.e. $\boldsymbol{\omega} = [0, 0, 1]^\top$. This assumption leads to the *eikonal equation*, which is a particular case arising from the differential SfS model (1.5):

$$|\nabla u(\mathbf{x})| = f(\mathbf{x}) := \sqrt{\frac{1}{I(\mathbf{x})^2} - 1}, \qquad (1.6)$$

where the graylevel function I, which typically takes integer values between 0 and 255, is implicitly resampled to take real values in the range [0, 1].

Note that even in the most simple case (the eikonal equation) we get a nonlinear PDE of the first order, and the solutions are a priori non-differentiable and non-unique, even if we complement Eq. (1.6) with a Dirichlet boundary condition, i.e. $u = g$ on the boundary $\partial\Omega$ of Ω. Moreover, the right-hand side of the eikonal equation is not always defined because $I(\mathbf{x})$ can vanish at some points. A simple example is the hemisphere $z = \sqrt{1 - x^2 - y^2}$ under a parallel frontal lighting. In this case, $I(\mathbf{x}) = 0$ at the equator. However, in that simple situation the boundary condition $u = 0$ can help to solve the problem.

Under oblique light direction the same example becomes more difficult because there will be a *black shadow region* $\Omega_s \subset \Omega$ where $I(\mathbf{x}) \equiv 0$, and in that region the model has no information to reconstruct the surface. The boundary of Ω_s, which is not known a priori, is a curve where it would be difficult to impose boundary conditions in the numerical approximation. In general, the curve separating the region Ω_s will depend on the shape of the surface and on the light source direction ω. Note that in case of black shadows, the model is clearly unable to produce a reasonable surface approximation, because the information is missing. In this situation, one can follow a global approach avoiding to impose boundary conditions on $\partial\Omega_s$. This leads to the concept of *maximal solution*, where we solve the PDE on the whole domain with the standard Dirichlet boundary condition on $\partial\Omega$ (not on $\partial\Omega_s$), and recover a linear reconstruction on Ω_s (we refer to [17, 34, 35] for more details).

1.2.3 Ill-Posedness of the SfS Models

The "workshop metaphor" illustrated in Fig. 1.2 is representative of the ill-posedness of SfS, because a posteriori estimating 3D-geometry from a single image is possible only if reflectance and lighting are known a priori. The reliability of these priors is of fundamental importance to guarantee that the solution of SfS is meaningful.

This is illustrated in Fig. 1.3, which shows how the assumptions leading to Eq. (1.6), i.e. a uniform reflectance ($r(\mathbf{x}) \equiv 1$) and a parallel uniform lighting ($\psi(\mathbf{x}) \equiv 1$ and $\omega = [0, 0, 1]^\top$), which is just a rough approximation in the case of Fig. 1.3a, yield the erroneous interpretation of the 3D-shape shown in Fig. 1.3b. However, this solution is an exact solution of Eq. (1.6), as shown by frontally relighting this uniformly white 3D-shape (cf. Fig. 1.3c).

Even when reflectance and lighting are known, i.e. when $r(\mathbf{x}) \equiv 1$ and $\psi(\mathbf{x}) \equiv 1$, the non-differential model (1.2) of SfS remains ill-posed:

$$I(\mathbf{x}) = \omega^\top \mathbf{n}(\mathbf{x}). \tag{1.7}$$

Except for some sparse *singular points*, where $\mathbf{n}(\mathbf{x})$ points in the same direction of ω, there exists an infinity of surface normals explaining the graylevel in one pixel. It

(a)   (b)   (c)

Fig. 1.3 Illustration of the importance of reflectance and lighting priors on the solution of SfS [25]. **a** A well-known graylevel image $I(\mathbf{x})$. **b** 3D-shape estimated by solving the SfS model (1.6), by (wrongly) assuming uniform reflectance $r(\mathbf{x}) \equiv 1$, and uniform frontal lighting, i.e. $\psi(\mathbf{x}) \equiv 1$ and $\boldsymbol{\omega} = [0, 0, 1]^{\top}$. The 3D-reconstruction largely departs from the true geometry. Yet, by taking from above a picture of the uniformly white 3D-shape **b**, using the camera's flash as single light source, one gets image **c**, which resembles **a**. 3D-shape **b** is thus a plausible explanation of **a**: the bias comes from the inappropriate reflectance and lighting priors

comes down from Eq. (1.7) that these normals $\mathbf{n}(\mathbf{x})$ form a revolution cone around the lighting direction $\boldsymbol{\omega}$. It is thus very difficult to *locally* solve SfS.

A simple example in dimension 1 is given by the surface $z = u_1(x) = 1 - x^2$ in the interval $(-1, 1)$ (cf. Fig. 1.4a), under vertical lighting $\boldsymbol{\omega} = [0, 1]^{\top}$, which satisfies an equation of the form (1.6) and the homogeneous boundary condition $u_1(1) = u_1(-1) = 0$. However, the function $u_2(x) = -u_1(x)$ in the same interval still satisfies this equation, since $|\nabla u(\mathbf{x})|$ is the same and the same boundary condition holds. This example is an illustration of the famous *concave/convex ambiguity* of SfS.

Note that u_1 and u_2 are two differentiable solutions to the same problem. If we decide to accept also solutions which are differentiable only almost everywhere (a very natural choice in view of real applications), we suddenly have an infinite number of solutions which can be obtained just considering all the possible reflections of one of those solutions, e.g. u_1, with respect to a horizontal axis located at the height $z = h$, where $h \in (0, 1)$. This is illustrated in Fig. 1.4b, where three such solutions are exhibited. If one fixes the height at the singular point, then only one of these solutions can be accepted (we refer to [64] for this result), but such an additional knowledge is clearly not very realistic. A general theory for weak solutions of Hamilton–Jacobi equations (that includes the eikonal equation) has been developed in the last 20 years starting from the seminal paper by Crandall and Lions [24]. We refer the interested reader to the book [8] and the references therein.

Practical ways to reduce this ambiguity include resorting to more realistic models such as perspective camera [117] or near-lighting [89]. However, it has been shown that this remains insufficient to ensure well-posedness [16]. Recently, the introduction of an attenuation factor in the brightness equations relative to various perspective SfS models allowed to make the corresponding differential problems well-posed. In [18], a unified approach based on the theory of *viscosity solutions* has been proposed, showing that the brightness equations coming from different non-

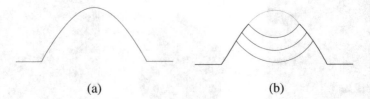

Fig. 1.4 a Example of the 1D-surface $z = u_1(x) = 1 - x^2$. **b** Under vertical lighting, three other solutions (amongst an infinity), which are differentiable almost everywhere, satisfy the same eikonal equation as u_1

Lambertian reflectance models with the attenuation term admit a unique viscosity solution.

1.2.4 Numerical Approximation

An important step towards the numerical solving of SfS was achieved when inverse problems in computer vision caught the attention of mathematicians [64]. Efficient numerical approaches were suggested to solve the eikonal equation (1.6), which is the SfS fundamental differential model relating the surface slope to the image graylevel. By construction, this equation can only be solved *globally*, therefore, SfS ambiguities are reduced, in comparison with local approaches. They are not eliminated yet, because the concave/convex ambiguity remains.

An overview of the numerical methods for solving SfS can be found in [31, 134]. PDE-based methods (e.g. [34]) find a viscosity solution to the eikonal equation. Just to give an example let us consider the basic eikonal equation (1.6). A typical technique to solve it is using a finite difference scheme. One example is the following iterative Lax–Friedrichs scheme which, in its simplest form, can be written as

$$
u_{i,j}^{(k+1)} = \frac{u_{i-1,j}^{(k)} + u_{i+1,j}^{(k)} + u_{i,j-1}^{(k)} + u_{i,j+1}^{(k)}}{4}
$$
$$
- \frac{1}{2} \left(\sqrt{\left(\frac{u_{i+1,j}^{(k)} - u_{i-1,j}^{(k)}}{2} \right)^2 + \left(\frac{u_{i,j+1}^{(k)} - u_{i,j-1}^{(k)}}{2} \right)^2} - f_{i,j} \right),
\tag{1.8}
$$

where $f_{i,j}$ is the right-hand side of the eikonal equation (1.6) at the pixel (i, j), $u_{i,j}$ is the height at this pixel, and the index k is the number of the iteration of the iterative scheme. The values $\{u_{i,j}^0\}$ represent an initial guess for the height, typically a constant value. Let us briefly explain the meaning of the iterative scheme: the first term is an average of four values around the pixel (i, j), and inside the square root there are the centered finite difference approximations of the partial derivatives $\partial u / \partial x$ and $\partial u / \partial y$. In practice, several approximation schemes are available, e.g.

finite difference as illustrated in [86, 103], semi-Lagrangian schemes [32, 33]. Most of the efficient schemes use upwind approximations of the derivatives and additional terms to control the diffusion in the scheme. Let us also mention that a fast-marching version for these methods allows to drastically reduce the CPU time for this type of algorithms [26, 103] and has been extensively applied in the area of image processing. Another delicate point is that the graylevel function I is typically a discontinuous function, so the approximation scheme should take into account this lack of regularity (a result in this direction is in [36]).

On the other hand, many optimisation-based methods have been proposed to compute the normal field $\mathbf{n}$. Under orthographic projection, (1.3) shows that $\mathbf{n}$ just depends on p and q as defined in (1.4). Therefore, there exists a function $\mathcal{R}$ such that $\mathcal{R}(p(\mathbf{x}), q(\mathbf{x})) := R(\mathbf{n}(\mathbf{x}))$. From this and Eq. (1.1), the following least-squares variational model of SfS is derived (robust estimators have also been used [130]):

$$\min_{p,q:\,\Omega\to\mathbb{R}} \int_{\Omega} \left| I(\mathbf{x}) - \mathcal{R}(p(\mathbf{x}), q(\mathbf{x})) \right|^2 d\mathbf{x}. \tag{1.9}$$

As already said in the previous subsection, this problem is clearly ill-posed. Nevertheless, if u is of class C^2, p and q are two non-independent functions since, according to Schwarz's theorem, $\partial p/\partial y = \partial q/\partial x$. For numerical reasons [49], this hard constraint is usually replaced by a quadratic regularisation term weighted by a hyper-parameter $\lambda > 0$, which gives the following better-posed problem than (1.9):

$$\min_{p,q:\,\Omega\to\mathbb{R}} \int_{\Omega} \left| I(\mathbf{x}) - \mathcal{R}(p(\mathbf{x}), q(\mathbf{x})) \right|^2 d\mathbf{x} + \lambda \int_{\Omega} \left| \frac{\partial p}{\partial y}(\mathbf{x}) - \frac{\partial q}{\partial x}(\mathbf{x}) \right|^2 d\mathbf{x}. \tag{1.10}$$

Another regularisation term has been extensively used, since it is easier to discretise [62]:

$$\min_{p,q:\,\Omega\to\mathbb{R}} \int_{\Omega} \left| I(\mathbf{x}) - \mathcal{R}(p(\mathbf{x}), q(\mathbf{x})) \right|^2 d\mathbf{x} + \lambda \int_{\Omega} \left[|\nabla p(\mathbf{x})|^2 + |\nabla q(\mathbf{x})|^2 \right] d\mathbf{x}. \tag{1.11}$$

Typical optimisation methods are descent methods. For instance, the Euler–Lagrange equations derived from (1.11) are written (dependencies on $\mathbf{x}$ are omitted):

$$\left[I - \mathcal{R}(p, q) \right] \frac{\partial \mathcal{R}}{\partial p}(p, q) + \lambda \,\Delta p = 0 \quad \text{and} \quad \left[I - \mathcal{R}(p, q) \right] \frac{\partial \mathcal{R}}{\partial q}(p, q) + \lambda \,\Delta q = 0. \tag{1.12}$$

Using the classical discrete approximation of the Laplacian Δp at pixel (i, j):

$$\Delta p_{i,j} \approx \frac{p_{i+1,j} + p_{i-1,j} + p_{i,j+1} + p_{i,j-1}}{4} - p_{i,j}, \tag{1.13}$$

the following iterative scheme for solving (1.11) comes down from (1.12) and (1.13) [53]:

$$\begin{cases} p_{i,j}^{(k+1)} = \overline{p}_{i,j}^{(k)} + \dfrac{1}{\lambda}\left[I_{i,j} - \mathcal{R}\left(p_{i,j}^{(k)}, q_{i,j}^{(k)}\right)\right]\dfrac{\partial \mathcal{R}}{\partial p}\left(p_{i,j}^{(k)}, q_{i,j}^{(k)}\right), \\[2ex] q_{i,j}^{(k+1)} = \overline{q}_{i,j}^{(k)} + \dfrac{1}{\lambda}\left[I_{i,j} - \mathcal{R}\left(p_{i,j}^{(k)}, q_{i,j}^{(k)}\right)\right]\dfrac{\partial \mathcal{R}}{\partial q}\left(p_{i,j}^{(k)}, q_{i,j}^{(k)}\right), \end{cases} \quad (1.14)$$

where $\overline{p}$ denotes the local average of p, $p^{(0)}$ and $q^{(0)}$ are given initial conditions, and the index k is the number of the iteration.

To avoid divergence for such schemes [30], it has been proposed to directly minimise the functional in (1.11), using conjugate gradient descent [62, 115] or line search [29], but the approximate solution is typically a local minimum. A way to overcome this limitation is to use a global optimisation, e.g. simulated annealing [27]. Finally, to decrease the CPU time, it has been dealt with multi-resolution [115].

Even if some optimisation-based methods aim to directly solve the SfS problem in the height u, as for instance [23] where a parametric model with few parameters is used, most of them first compute a normal field $\mathbf{n}$. Once the components p and q of the normal (cf. Eq. (1.4)) have been computed, it remains to *integrate* them into a height map. Several methods can be used for this task, depending on the application's requirements in terms of speed, robustness to noise in the estimated normal field and preservation of discontinuities [95]. For instance, a standard solution for the recovery of a smooth height map consists in considering the quadratic variational problem:

$$\min_{u:\Omega\to\mathbb{R}} \int_{\Omega} \left| \nabla u(\mathbf{x}) - \begin{bmatrix} p(\mathbf{x}) \\ q(\mathbf{x}) \end{bmatrix} \right|^2 \mathrm{d}\mathbf{x}, \quad (1.15)$$

which can be solved, e.g. using Fourier analysis [37], discrete sine or cosine transform [109] or iterative methods [7], depending upon the shape of Ω and the boundary conditions.

1.2.5   Applications of SfS

The natural application of SfS is the 3D-reconstruction of a scene from a single image. However, in real-world settings the assumptions formulated above on reflectance and lighting are too restrictive. Therefore, efforts have recently been devoted to move beyond the assumptions of Lambertian reflectance [56, 118, 119] and controlled illumination [55, 92]. In such works, reflectance and lighting are allowed to take a more general form, yet they still must be calibrated. To remove this limitation, additional priors must be introduced, as it is common in the field of intrinsic image decomposition where the reflectance is often assumed to be piecewise smooth [9]. Alternatively, deep learning techniques can be employed to simultaneously estimate shape, reflectance and lighting, provided that the object to reconstruct resembles those in the learning database [102]. In the absence of such priors, SfS can be combined with another 3D-reconstruction technique: the latter provides a coarse prior on geometry, whose details are then refined using SfS. In this view, SfS has been combined with shape-from-texture [123], structure-from-motion [38], multi-view stereopsis [61, 68, 72] or depth sensors [41, 85].

An alternative strategy to resolve the ambiguities of SfS consists in using additional images taken under varying lighting. This approach, which is called photometric stereo, will be discussed in the next section.

1.3 Photometric Stereo

The photometric stereo technique, first developed by Woodham [129], is an extension of SfS which considers several images acquired under the same viewing angle, but various lighting conditions.

1.3.1 Well-Posedness of PS

One may reasonably hope that shape inference by PS will be better-posed, in comparison with the single-image case of SfS. Indeed, 3D-shape and Lambertian reflectance can be exactly and uniquely determined from a set of three images taken under non-coplanar, uniform, calibrated directional lighting. This is easily shown by considering a system of $m \geq 3$ image irradiance equations such as (1.2), obtained under illumination with uniform intensity $\psi(\mathbf{x}) \equiv 1$, but varying direction characterised by vectors ω_i, $i \in \{1, \ldots, m\}$:

$$I_i(\mathbf{x}) = r(\mathbf{x}) \, \omega_i^\top \mathbf{n}(\mathbf{x}), \qquad i \in \{1, \ldots, m\}. \tag{1.16}$$

This system of equations comes down to a linear system of m equations in $\mathbf{m}(\mathbf{x}) := r(\mathbf{x}) \, \mathbf{n}(\mathbf{x})$. Provided that $m = 3$ and the three illumination vectors ω_i, $i \in \{1, 2, 3\}$, are non-coplanar, there exists a unique solution $\mathbf{m}(\mathbf{x})$ of this system, from which the albedo can be extracted as $r(\mathbf{x}) = |\mathbf{m}(\mathbf{x})|$ and the surface normal as $\mathbf{n}(\mathbf{x}) = \frac{\mathbf{m}(\mathbf{x})}{|\mathbf{m}(\mathbf{x})|}$. When $m > 3$, an approximate solution of the system can be estimated as long as the m illumination vectors remain non-coplanar. An example of result obtained with this approach on a banknote is presented in Fig. 1.5. It illustrates well the unique ability of PS both to estimate fine-scale geometric details, and to estimate the reflectance.

PS can however be ill-posed in two particular scenarios. Firstly, when lighting is unknown (*uncalibrated PS*), the local estimation of surface normals is underconstrained. As in SfS, the problem must be reformulated globally, and the integrability constraint must be imposed [133]. But even then, a low-frequency ambiguity known as the *generalised bas-relief ambiguity* remains [12]: it is necessary to introduce additional priors, see [107] for an overview of existing uncalibrated photometric stereo approaches, and [19] for a modern solution based on deep learning. Another situation where PS is ill-posed is when only two images are considered [91]: in each

Fig. 1.5 Photometric stereo-based 3D-reconstruction of a 10 euro banknote. From a set of images captured under varying lighting (left), PS infers both the surface geometry (top-right, we show the RGB-coded estimated normals and the 3D-shape obtained by integration of the normals), as well as its reflectance (bottom-right, we show the estimated albedo)

pixel there exist two possible normals explaining the pair of graylevels, even with known reflectance and lighting. Again, integrability must be imposed in order to limit the ambiguities [84].

1.3.2 Numerical Solving of PS

In the previous subsection, we described a simple strategy to estimate the surface normals by photometric stereo. The knowledge of surface normals is however not sufficient to fully characterise the geometry of the pictured scene. To obtain a complete 3D-representation, the normals must then be integrated into a height map. We have already discussed this integration problem in Sect. 1.2.4, and we refer the reader to [95] for a comprehensive overview.

With this pipeline, one first estimates the surface normals, and then integrates them into a height map. This strategy is however suboptimal, since any error in the normal estimation step will propagate during the subsequent normal integration one. An alternative strategy is to reformulate (1.16) as a system of partial differential equations in the unknown height map u, and directly estimate u. For instance, one may consider the ratio of two equations such as (1.16), for $i \neq j$, while replacing the surface normal $\mathbf{n}(\mathbf{x})$ by its definition (1.3). This yields the following PDE:

$$\left[I_i(\mathbf{x})\,\boldsymbol{\omega}_j - I_j(\mathbf{x})\,\boldsymbol{\omega}_i \right]^\top \begin{bmatrix} -\nabla u(\mathbf{x}) \\ 1 \end{bmatrix} = 0, \qquad \forall \mathbf{x} \in \Omega, \qquad (1.17)$$

which is linear in ∇u, and is independent from the reflectance r. It can be solved, e.g. using a finite difference upwind scheme or semi-Lagrangian methods [69]. When more than a single pair of images is considered, the joint approximate solving of the system of equations such as (1.17), obtained for every pair $\{i, j\}$, can be formulated as a variational problem:

$$\min_{u: \Omega \to \mathbb{R}} \sum_{i<j} \sum \int_{\Omega} \left| \left[I_i(\mathbf{x}) \, \omega_j - I_j(\mathbf{x}) \, \omega_i \right]^{\top} \begin{bmatrix} -\nabla u(\mathbf{x}) \\ 1 \end{bmatrix} \right|^2 d\mathbf{x}. \qquad (1.18)$$

Such an approach, initially proposed in [90, 110], also easily extends to more elaborate camera or reflectance models [71].

Nevertheless, this ratio-based approach does not provide the reflectance, contrarily to the simple pipeline presented in the previous subsection. Moreover, solving the linearised partial differential equations (1.17) is not equivalent to solving the original Eqs. (1.16): for instance, Gaussian noise on the images turns into Cauchy noise on the ratios, making least-squares inference suboptimal. Thus, the joint recovery of height and reflectance by variational inversion of the image irradiance Eqs. (1.16) has also been explored. For example, plugging the definition (1.3) of $\mathbf{n}(\mathbf{x})$ into (1.16), the joint estimation of height and reflectance in a least-squares sense leads to

$$\min_{u, r: \Omega \to \mathbb{R}} \sum_{i=1}^{m} \int_{\Omega} \left| I_i(\mathbf{x}) - r(\mathbf{x}) \, \omega_i^{\top} \begin{bmatrix} -\nabla u(\mathbf{x}) / \sqrt{1 + |\nabla u(\mathbf{x})|^2} \\ 1 / \sqrt{1 + |\nabla u(\mathbf{x})|^2} \end{bmatrix} \right|^2 d\mathbf{x}, \qquad (1.19)$$

which can be solved, e.g. using alternating reweighted least-squares [93].

1.3.3   PS with Non-trivial Reflectance or Lighting

The surface has been assumed Lambertian in our models, and lighting has been assumed directional but those assumptions are difficult to satisfy in real-world scenarios. An important feature of PS, in comparison with SfS, is that the redundancy provided by the multiple images enables relaxing such assumptions. Indeed, shadows or off-Lambertian effects such as specularities can be coped with by solving PS in a robust manner, for instance, by resorting to sparse regression which treats such effects as outliers to the Lambertian model [52, 93]. Other ways to deal with off-Lambertian effects include inverting a reflectance model which is more sophisticated than Lambert's [71, 106] or pre-processing the images according to a low-rank prior [131]. Let us also mention data-driven methods, which either compare the intensity variations with those observed on a reference object with known shape [46] or resort to a deep neural network trained or large dataset [98].

Another direction of research on PS is the study of more realistic lighting models, in order to simplify the acquisition of data. For instance, some methods have been developed to handle images acquired under nearby point light illumination [70], which finds a natural application in LED-based photometric stereo [94]. This permits to build a simple acquisition setup based on cheap hardware. Extended light sources have also been considered, which permits for instance to use the screen of a LCD display as light source [21]. Eventually, other approaches have considered the case of natural illumination in order to bring PS outdoor [11], and numerical solving methods based on variational principles [42] or deep learning [47] have recently been suggested.

1.3.4 Combining PS and Other 3D-Reconstruction Methods

A criticism which is frequently formulated against PS is that it excels with the recovery of high-frequency geometric details, yet it is prone to a low-frequency bias which may distort the global geometry [82]. In fact, such a bias usually comes from a contradiction between the assumptions behind the image formation model and the actual experiments, e.g. assuming a directional light source instead of a nearby point light one. Therefore, the methods discussed in the previous subsection provide a remedy to such a bias.

On the other hand, it is sometimes simpler from a practical perspective to stick to the simplest assumptions, and rather remove the low-frequency bias by coupling PS with another 3D-reconstruction method such as shape-from-silhouette [122], multi-view stereopsis [63] or depth sensing [87]. In such works, PS provides the fine-scale geometric details, which are combined with the gross geometry provided by the alternative technique.

Another interesting application of PS is 3D-reconstruction from a single shot, which can be achieved by using a multichannel camera coupled with monochromatic coloured light sources which are simultaneously turned on: each channel can then be viewed as a graylevel image obtained under a single light source. This idea, which dates back from the nineties [60], has more recently been applied to the real-time 3D-reconstruction of deformable surfaces by combining PS with optical flow [44] or scene flow [40].

1.3.5 Applications of PS

The ability of PS to estimate both the fine-scale geometric details and the reflectance of the surface has proven useful in many applications. Here, we briefly highlight a few of them. For instance, PS can be used to infer 3D-models for augmented reality, which can be very helpful for computer-aided surgery using laparoscopy [22]. Another medical application of PS is the characterisation of the melanoma's shape and color, as proposed in [114]. Besides medical applications, PS has been extensively used in the field of quality control, e.g. for the inspection of defects on metallic surfaces [113]. Also, let us mention Reflectance Transform Imaging (RTI) techniques, which are based on PS principles, and allow one to interactively relight the pictured surfaces. Such an approach finds a natural application in the field of cultural heritage, see the recent survey [88] for an overview. Finally, Chap. 7 in the present volume addresses a novel application, which is the estimation of facial aging.

1.4 Shape-from-Polarisation

Another problem belonging to the shape-from-X class is the shape-from-polarisation one. The goal is the same, i.e. recover the 3D-shape of the object, but starting from a different input data, given by polarisation information.

1.4.1 Description and Generation of a Polarisation Image

When unpolarised light is reflected by a surface, it becomes partially polarised [128]. This applies to both specular [96] and diffuse [4] reflections caused by subsurface scattering. Using a linear polarising filter placed in front of a camera, a sequence of $m \geq 3$ images (cf. Fig. 1.6a) is captured by rotating the filter under varying polariser angle ϑ_j, $j \in \{1, \ldots, m\}$. The measured brightness at each pixel $\mathbf{x}$ varies in accordance to the *transmitted radiance sinusoid* corresponding to

$$i_{\vartheta_j}(\mathbf{x}) = \frac{I_{max}(\mathbf{x}) + I_{min}(\mathbf{x})}{2} + \frac{I_{max}(\mathbf{x}) - I_{min}(\mathbf{x})}{2} \cos[2\vartheta_j - 2\phi(\mathbf{x})], \quad (1.20)$$

where $\phi(\mathbf{x})$ is the phase angle, I_{max} the maximum measured pixel brightness and I_{min} the minimum one.

A *polarisation image* (cf. Fig. 1.6b–d), i.e. the full set of polarisation data for a given object or scene, can be obtained by decomposing the sinusoid at every pixel into three separate components [127]: the *phase angle*, $\phi(\mathbf{x})$, the *unpolarised intensity*, $i_{un}(\mathbf{x})$ and the *degree of polarisation*, $\rho(\mathbf{x})$, where

$$i_{un}(\mathbf{x}) = \frac{I_{max}(\mathbf{x}) + I_{min}(\mathbf{x})}{2} \quad \text{and} \quad \rho(\mathbf{x}) = \frac{I_{max}(\mathbf{x}) - I_{min}(\mathbf{x})}{I_{max}(\mathbf{x}) + I_{min}(\mathbf{x})}. \quad (1.21)$$

The phase angle $\phi(\mathbf{x})$ is directly related to the angle of the linearly polarised component of the reflected light and can be defined as the angle of maximum or minimum transmission. Since polarisers cannot distinguish between two angles separated by π radians, the range of initially acquired phase measurements is $[0, \pi)$. Therefore, there is a π ambiguity, since two maxima in pixel brightness are found as the polariser is rotated through 2π. The unpolarised image $i_{un}(\mathbf{x})$ is simply the image that would be obtained using a standard camera. The degree of polarisation $\rho(\mathbf{x})$ can be defined in terms of refractive index and zenith angle of the surface normal [128], but the

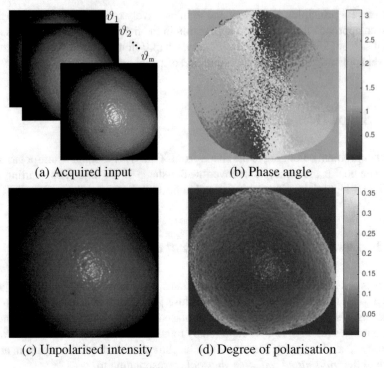

(a) Acquired input   (b) Phase angle

(c) Unpolarised intensity   (d) Degree of polarisation

Fig. 1.6 **a** Polarimetric capture, and **b–d** decomposition into polarisation images, from captured data of a piece of fruit. Pictures taken and adapted from [112]

explicit formula is different depending on the polarisation model used, as we will see in Sect. 1.4.2 below.

These quantities can be estimated from the captured image sequence using different methods, e.g. the Levenberg–Marquardt nonlinear curve fitting algorithm [4], linear methods [50] or following the procedure suggested by Wolff in [127] for the specific case of $m = 3$, $\vartheta \in \{0, \frac{\pi}{4}, \frac{\pi}{2}\}$.

1.4.2 Diffuse and Specular Polarisation Models

A polarisation image provides information on the azimuth and zenith angles of the normal, and, hence, a constraint on the surface normal direction at each pixel. The exact nature of the constraint depends on the polarisation model used.

Using a diffuse polarisation model, the phase angle $\phi(\mathbf{x})$ is the polariser angle ϑ_j at which $I_{\max}$ is observed. It determines the azimuth angle $\alpha(\mathbf{x}) \in [0, 2\pi[$ of the surface normal up to a π ambiguity: $\alpha(\mathbf{x}) = \phi(\mathbf{x})$ or $\phi(\mathbf{x}) + \pi$. The degree of polarisation $\rho_d(\mathbf{x})$, on the other hand, is related to the zenith angle $\theta(\mathbf{x}) \in [0, \frac{\pi}{2}]$ of the normal in viewer-centered coordinates (i.e. the angle between the normal and viewer) as follows:

$$\rho_d(\mathbf{x}) = \frac{\sin^2 \theta(\mathbf{x}) \left(\eta - \frac{1}{\eta}\right)^2}{4 \cos \theta(\mathbf{x}) \sqrt{\eta^2 - \sin^2 \theta(\mathbf{x})} - \sin^2 \theta(\mathbf{x}) \left(\eta + \frac{1}{\eta}\right)^2 + 2\eta^2 + 2}, \quad (1.22)$$

where η is the refractive index (in general, η is unknown, but for most dielectrics typical values range between 1.4 and 1.6, hence an accurate estimate of geometry can be obtained without a precise estimate of η [4]).

Instead, using a specular polarisation model, the azimuth angle of the surface normal is perpendicular to the phase of the specular polarisation [97] leading to a $\frac{\pi}{2}$ shift, so that the azimuth angle corresponds to polariser angle ϑ_j at which $I_{\min}$ is observed: $\alpha(\mathbf{x}) = \phi(\mathbf{x}) \pm \frac{\pi}{2}$. Regarding the degree of polarisation $\rho_s(\mathbf{x})$, it relates to the zenith angle according to

$$\rho_s(\mathbf{x}) = \frac{2 \sin^2 \theta(\mathbf{x}) \cos \theta(\mathbf{x}) \sqrt{\eta^2 - \sin^2 \theta(\mathbf{x})}}{\eta^2 - \sin^2 \theta(\mathbf{x}) - \eta^2 \sin^2 \theta(\mathbf{x}) + 2 \sin^4 \theta(\mathbf{x})}, \quad (1.23)$$

and in that case the dependency of the degree of polarisation ρ_s on η is weaker than in the diffuse case.

1.4.3   3D-Shape Recovery Using Polarisation Information

The phase angle $\phi(\mathbf{x})$ (cf. Fig. 1.6b) and the degree of polarisation $\rho(\mathbf{x})$ (cf. Fig. 1.6d) of reflected light convey information about the surface orientation through information on zenith and azimuth angles and, therefore, provide a cue for 3D-shape recovery.

There are nice and attractive properties to the SfP cue: it requires only a single viewpoint and a single illumination condition, it is invariant to illumination direction and surface albedo, and it provides information about both the zenith and azimuth angle of the surface normal. Unfortunately, the polarisation information alone restricts the surface normal at each pixel to two possible directions, providing in such a way only ambiguous estimates of the surface orientation.

SfP methods can be categorised into three groups:

1. Methods which use only polarisation information (cf. Sect. 1.4.3.1). They are passive since, typically, a polarisation image is obtained by capturing a sequence of images in which a linear polarising filter is rotated in front of the camera (possibly with unknown rotation angles [100]). These methods can be considered "single shot" methods by using custom CCD cameras configured for polarisation

imaging[1] or by mounting the polarisation filter on a CMOS sensor in order to acquire polarisation information in real time[2]).
2. Methods which combine polarisation with shading cues (cf. Sect. 1.4.3.2).
3. Methods which combine a polarisation image with an additional cue (cf. Sect. 1.4.3.3) such as stereo, multispectral measurements, an RGBD sensor or active polarised illumination.

SfP methods can also be classified according to the polarisation model (dielectric versus metal, diffuse [4, 50, 77], specular [79] or hybrid models [116]) and whether they compute shape in the surface normal or surface height domain.

1.4.3.1 Resolution Using Only Polarisation Information

The earliest work focused on capture, decomposition and visualisation of polarisation images was by Wolff in the nineties [127], even if older works on shape recovery by polarisation information exist since 1962 [108]. Both Atkinson and Hancock [4] and Miyazaki et al. [77] disambiguated the polarisation normals via propagation from the boundary under an assumption of global convexity. Huynh et al. [50] also disambiguated polarisation normals with a global convexity assumption, estimating the refractive index in addition. These works used a diffuse polarisation model whereas Morel et al. [79] used a specular polarisation model for metals. Recently, Taamazyan et al. [116] introduced a mixed diffuse/specular polarisation model. All of these methods estimate surface normals which must then be integrated into a height map. Moreover, since they rely entirely on the weak shape cue provided by polarisation and do not enforce integrability, the results are extremely sensitive to noise.

1.4.3.2 Polarisation and Shading Cues

A polarisation image also contains an unpolarised intensity channel (cf. Fig. 1.6c), which provides a shading cue. Mahmoud et al. [67] used a shape-from-shading cue assuming known light source direction, known albedo and Lambertian reflectance, in order to disambiguate the polarisation normals. Atkinson and Hancock [6] used calibrated, three-source Lambertian photometric stereo for disambiguation but avoiding an assumption of known albedo. Smith et al. [111] showed how to express polarisation and shading constraints directly in terms of surface height, leading to a robust and efficient linear least-squares solution. They also showed how to estimate the illumination, up to a binary ambiguity, making the method uncalibrated. However, they require known or uniform albedo. This requirement was afterwards relaxed in [112], where spatially varying albedo was estimated from a single polarisation image, assuming known illumination and strong smoothness assumptions. In [120]

---

[1]http://www.fluxdata.com/products/fd-1665p-imaging-polarimeter.
[2]https://www.ricoh.com/technology/tech/051_polarization.

variants of the aforementioned method have been exploited by introducing additional constraints which arise when a second light source is considered, allowing to relax the uniform albedo assumption even under unknown lighting. In this work, albedo-invariant or phase-invariant formulations were proposed. Another differential approach has been proposed in [65], where the geometry of the object is described through its level-sets for both diffuse and specular reflections. Ngo et al. [83] derived constraints which allowed surface normals, light directions and refractive index to be estimated from polarisation images under varying lighting. However, this approach requires at least four light sources.

1.4.3.3   Combining Polarisation with Other Cues

In order to solve the ambiguities generated by models using only polarisation information, some attempts have been done combining SfP with other cues. In addition to photometric cues (from SfS or PS), auxiliary geometric information can be considered. Stereo cues has been combined with polarisation to obtain surface orientation information since the nineties [126]. Rahmann et al. [96] proposed to reconstruct specular surfaces taking polarisation images from multiple views. The reconstruction problem is solved by an optimisation scheme where the surface geometry is modelled by a set of hierarchical basis functions. Atkinson et al. [3, 5] refined estimates of the surface normal to establish correspondences between two views of an object, extracting surface patches from each view. Multi-View Stereo (MVS) and polarisation have also been adopted for transparent and specular objects [73, 76], and a polarimetric MVS method applied to objects with mixed polarisation models is proposed in [28]. With respect to this last paper, which is offline and needs a manual preparation, Yang et al. proposed in [132] a fully automatic approach to produce a height map in real time using two views. More than two views have been used in [20]. Space carving [75, 76] or RGBD sensors [57, 58] have been employed to obtain initial 3D-shape, from which the ambiguities in SfP are resolved. Zhu et al. [135] used polarisation and an RGBD stereo pair to disambiguate the polarisation surface normal estimates using a higher order graphical model. Cameras with multiple spectral bands [51, 74] could be useful for disambiguating and estimating the refractive index of the surface.

1.4.4   An Example of Numerical Resolution for Shape Recovery

In this section, we want to give an example of numerical resolution of the SfP problem, either by following a non-differential approach, which considers as unknowns the partial derivatives p and q as defined in Eq. (1.4), or by solving a linear differential system directly in the height u.

We assume orthographic projection and directional illumination. We consider only the diffuse polarisation model, hence, the degree of polarisation is defined as in (1.22), and the object we want to recover is composed by dielectric (i.e. non metallic) materials. Moreover, the refractive index η is supposed to be a known constant, and interreflections are neglected. In order to estimate the phase angle $\phi(\mathbf{x})$ and the degree of diffuse polarisation $\rho_d(\mathbf{x})$ at each point, we fit the data to the transmitted radiance sinusoid (1.20) following one of the aforementioned methods, e.g. the idea by Wolff [127]. The zenith angle $\theta(\mathbf{x})$ of the surface normal can be obtained from Eq. (1.22) arriving to

$$\cos\theta(\mathbf{x}) = \mathbf{n}(\mathbf{x}) \cdot \mathbf{v} = f(\rho(\mathbf{x}), \eta) = \tag{1.24}$$

$$\sqrt{\frac{2\rho + 2\eta^2\rho - 2\eta^2 + \eta^4 + \rho^2 + 4\eta^2\rho^2 - \eta^4\rho^2 - 4\eta^3\rho\sqrt{-(\rho-1)(\rho+1)} + 1}{\eta^4\rho^2 + 2\eta^4\rho + \eta^4 + 6\eta^2\rho^2 + 4\eta^2\rho - 2\eta^2 + \rho^2 + 2\rho + 1}},$$

where we have denoted ρ_d simply by ρ and we have dropped the dependency of ρ on $\mathbf{x}$ for readability. The normal vector defined in (1.3) can be written in terms of azimuth and zenith angles as

$$\mathbf{n}(\mathbf{x}) = \begin{bmatrix} \cos\alpha(\mathbf{x})\sin\theta(\mathbf{x}) \\ \sin\alpha(\mathbf{x})\sin\theta(\mathbf{x}) \\ \cos\theta(\mathbf{x}) \end{bmatrix}. \tag{1.25}$$

Remembering that the phase angle $\phi(\mathbf{x})$ determines the azimuth angle $\alpha(\mathbf{x})$ of the normal up to a π ambiguity ($\alpha(\mathbf{x}) = \phi(\mathbf{x})$ or $\alpha = \phi(\mathbf{x}) + \pi$), the normal vector can be estimated up to an ambiguity. Several attempts have been done in order to disambiguate the azimuth angle, as explained in Sect. 1.4.3.1. Once the surface normal has been estimated, by integration we can recover the height, which is our real and final unknown to be found. Again, we refer the interested reader to Sect. 1.2.4 and the survey [95] for some discussion on the integration problem.

As an alternative, we can solve the problem directly in the unknown height following a differential approach, starting again from a single polarisation image, but using also the unpolarised intensity quantity, which is the image obtained using a standard camera for the SfS problem. For example, let us assume Lambertian reflectance, known illumination and uniform albedo that is factored into the light source vector $\boldsymbol{\omega}$. The shading constraint coming from the unpolarised intensity channel of a polarisation image reads as (cf. Eq. (1.5)):

$$i_{\mathrm{un}}(\mathbf{x}) = \frac{-\omega_1\, p(\mathbf{x}) - \omega_2\, q(\mathbf{x}) + \omega_3}{\sqrt{1 + p(\mathbf{x})^2 + q(\mathbf{x})^2}}. \tag{1.26}$$

Since we are working in a viewer-centered coordinate system, with the viewer $\mathbf{v} = [0, 0, 1]^\top$, Eq. (1.24) simplifies to $n_3(\mathbf{x}) = f(\rho(\mathbf{x}), \eta)$, which can be expressed in terms of the surface gradient as

$$f(\rho(\mathbf{x}), \eta) = \frac{1}{\sqrt{1 + p(\mathbf{x})^2 + q(\mathbf{x})^2}}. \tag{1.27}$$

Now, by using the image ratio technique commonly applied also in PS-SfS problems [119], taking a ratio between (1.26) and (1.27), the nonlinear normalisation factor vanishes, yielding the following linear equation in the surface gradient:

$$\frac{i_{un}(\mathbf{x})}{f(\rho(\mathbf{x}), \eta)} = -\omega_1 \, p(\mathbf{x}) - \omega_2 \, q(\mathbf{x}) + \omega_3. \tag{1.28}$$

Instead of disambiguating the polarisation normals at each pixel locally, as illustrated before following a non-differential approach, here we express the azimuth ambiguity as a collinearity condition which is satisfied by either of the two possible azimuth angles. In this way, we postpone resolution of the ambiguity until surface height is computed, solving the azimuthal ambiguities in a globally optimal way.

More in detail, for the diffuse case we require that the projection of the surface normal into the image plane Oxy, $[n_1(\mathbf{x}), n_2(\mathbf{x})]^\top$, is collinear with a vector pointing in the phase angle direction, $[\sin \phi(\mathbf{x}), \cos \phi(\mathbf{x})]^\top$. This requirement translates into the following condition:

$$\mathbf{n}(\mathbf{x})^\top [\cos \phi(\mathbf{x}), -\sin \phi(\mathbf{x}), 0]^\top = 0. \tag{1.29}$$

By rewriting $\mathbf{n}(\mathbf{x})$ in terms of the surface gradient, noting that the nonlinear normalisation term is always non-null, we obtain from Eq. (1.29) a second linear equation in the surface gradient:

$$- p(\mathbf{x}) \cos \phi(\mathbf{x}) + q(\mathbf{x}) \sin \phi(\mathbf{x}) = 0. \tag{1.30}$$

At this point, after approximating the surface gradient, e.g. by using finite differences, we arrive to a linear system of equations in terms of the unknown surface height, which can be solved using linear least-squares. For stability reasons, priors on convexity and smoothness can be added to the linear system. For more information on this idea and for details on the implementation, we refer the interested reader to [111, 112].

1.4.5   Applications

The polarisation state of light reflected by a surface provides a cue on the material properties of the surface and, via a relationship with surface orientation, the 3D-shape. Polarisation has been used for several applications since the nineties, including early work on material segmentation [128] and diffuse/specular reflectance separation [80]. In recent years, there has been an increasing interest in using polarisation information for 3D-shape estimation [57, 83, 111, 116]. Nice applications include polarised laparoscopy [45] or in general biomedical applications [121]. In addition to the use of polarisation information for 3D-reconstruction, recently several other applications are using polarisation for different tasks. For example, for image segmentation [104], robot dynamic navigation [13, 14], image enhancement [99, 100] and reflection separation by a deep learning approach [66], which simplifies previous works requiring three images from different polariser angles [59, 101, 124]. For more details on possible applications, we refer the interested reader to Chap. 6 of this volume.

1.5   A Short Presentation of This Volume

As we said in the introduction, the volume contains several contributions which represent recent trends in 3D-reconstruction via photometric techniques. Here is an overview of the chapters.

In Chap. 2, Breuss and Yarahmadi focus on a more realistic shape-from-shading model than that we described in Sect. 1.2, where perspective projection is considered. A comprehensive state of the art of perspective SfS (PSfS) is carried out. The case of a Lambertian surface illuminated either by a parallel and uniform luminous flux, or by a nearby point light source, is more specifically addressed. Finally, a comparative study is carried out between two methods of resolution of the PSfS problem under directional lighting, both of which are based on the fast-marching algorithm.

In Chap. 3, Or-El et al. tackle the problem of refining the depth map provided by RGBD sensors, by applying shape-from-shading techniques. The authors propose three ways to solve this problem. First, by a model-based approach effective for Lambertian surfaces, which refines the depth map by a SfS strategy applied to the RGB image. Then, they extend this approach to specular objects, using a Phong-type model and the InfraRed image with the attached (near) light source. Lastly, a deep learning-based solution is proposed.

In Chap. 4, Gallardo et al. tackle the problem of the 3D-reconstruction of deformable surfaces using non-rigid structure-from-motion and shading. The authors propose an optimisation-based strategy, which aims at finding the geometry (parameterised by vertices) and reflectance (parameterised by a finite set of albedo values) which minimise a cost function combining a shape-from-shading term and a structure-from-motion one. Additional terms are also included in the cost function: a contour boundary one, a smoothness one and a quasi-isometry one. The resulting non-convex optimisation problem is addressed by a careful heuristical initialisation followed by an iterative, Gauss–Newton-based refinement over all variables in a multi-scale fashion. The proposition is evaluated both qualitatively and quantitatively against the state of the art.

In Chap. 5, Brahimi et al. present a theoretical contribution on the well-posedness of uncalibrated photometric stereo under general illumination. In particular, they prove that there is no ambiguity for the perspective model if lighting is represented by first-order spherical harmonics. In the process of establishing their main result they also provide a comprehensive survey of the available results regarding the well-posedness of several photometric stereo problems and they examine in detail the case of the orthographic projection. For this problem they prove that, even in the case of spherical harmonics, the concave/convex ambiguity still persists. They conclude with some numerical experiments.

Chapter 6, authored by Shi et al., represents a concise survey on SfP. After an introduction, the authors briefly recall the Fresnel theory, which is the theoretical basis of polarisation imaging. The process for the formation of a polarisation image is described, giving also details on the data acquisition. The authors discuss the estimation of azimuth and zenith angles of the normal for surfaces with different reflectance properties (specular, diffuse, and mixed polarisation). Then, the combination of SfP with auxiliary information is explored, e.g. geometric cues, spectral cues, photometric cues and deep learning. Moreover, applications which can benefit from polarisation information, in addition to the 3D-shape recovery, are presented. The chapter ends with a discussion on problems still open.

Finally, Chap. 7 by Dahlan et al. addresses the problem of facial aging estimation, using light scattering photometry. It is shown that the roughness parameter of several BRDF models is correlated with the age. Therefore, facial aging estimation can be carried out by fitting a BRDF model to an input image. In this work, geometry estimation is carried out using photometric stereo, by resorting to an illumination dome. Then, given the estimated normals, an image with frontal lighting is used to infer the BRDF parameters. Various experiments are carried out to study whether these estimated parameters correlate with age and it is shown that this is the case for the roughness parameter. Several tests on real images are illustrated and analysed.

References

1. Adelson EH, Pentland AP (1996) The perception of shading and reflectance. In: Perception as Bayesian inference. Cambridge University Press, Cambridge, pp 409–423
2. AliceVision. https://alicevision.org/
3. Atkinson GA, Hancock ER (2005) Multi-view surface reconstruction using polarization. In: Proceedings of the IEEE international conference on computer vision 1:309–316
4. Atkinson GA, Hancock ER (2006) Recovery of surface orientation from diffuse polarization. IEEE Trans Image Process 15(6):1653–1664
5. Atkinson GA, Hancock ER (2007) Shape estimation using polarization and shading from two views. IEEE Trans Pattern Anal Mach Intell 29(11):2001–2017
6. Atkinson GA, Hancock ER (2007) Surface reconstruction using polarization and photometric stereo. In: Proceedings of the international conference on computer analysis of images and patterns, pp 466–473
7. Bähr M, Breuß M, Quéau Y, Boroujerdi AS, Durou J-D (2017) Fast and accurate surface normal integration on non-rectangular domains. Comput Vis Media 3(2):107–129
8. Barles G (1994) Solutions de viscosité des équations de Hamilton-Jacobi. Mathématiques et Applications, vol 17. Springer, Berlin
9. Barron JT, Malik J (2015) Shape, illumination, and reflectance from shading. IEEE Trans Pattern Anal Mach Intell 37(8):1670–1687
10. Bartoli A, Gérard Y, Chadebecq F, Collins T, Pizarro D (2015) Shape-from-template. IEEE Trans Pattern Anal Mach Intell 37(10):2099–2118
11. Basri R, Jacobs D, Kemelmacher I (2007) Photometric stereo with general, unknown lighting. Int J Comput Vis 72(3):239–257

12. Belhumeur PN, Kriegman DJ, Yuille AL (1999) The Bas-relief ambiguity. Int J Comput Vis 35(1):33–44
13. Berger K, Voorhies R, Matthies L (2016) Incorporating polarization in stereo vision-based 3D perception of non-Lambertian scenes. In: Unmanned systems technology XVIII. Proceedings of the SPIE, vol 9837, p. 98370P
14. Berger K, Voorhies R, Matthies LH (2017) Depth from stereo polarization in specular scenes for urban robotics. In: Proceedings of the international conference on robotics and automation, pp 1966–1973
15. Brady M, Yuille AL (1984) An extremum principle for shape from contour. IEEE Trans Pattern Anal Mach Intell 6(3):288–301
16. Breuß M, Cristiani E, Durou J-D, Falcone M, Vogel O (2012) Perspective shape from shading: ambiguity analysis and numerical approximations. SIAM J Imaging Sci 5(1):311–342
17. Camilli F, Grüne L (2000) Numerical approximation of the maximal solutions for a class of degenerate Hamilton-Jacobi equations. SIAM J Numer Anal 38(5):1540–1560
18. Camilli F, Tozza S (2017) A unified approach to the well-posedness of some non-Lambertian models in shape-from-shading theory. SIAM J Imaging Sci 10(1):26–46
19. Chen G, Han K, Shi B, Matsushita Y, Wong K-YK (2019) Self-calibrating deep photometric stereo networks. In: Proceedings of the IEEE conference on computer vision and pattern recognition, pp 8739–8747
20. Chen L, Zheng Y, Subpa-Asa A, Sato I (2018) Polarimetric three-view geometry. In: Proceedings of the European conference on computer vision, pp 20–36
21. Clark JJ (2010) Photometric stereo using LCD displays. Image Vis Comput 28(4):704–714
22. Collins T, Bartoli A (2012) 3D-reconstruction in laparoscopy with close-range photometric stereo. In: International conference on medical image computing and computer-assisted intervention. Lecture notes in computer science, vol 7511, pp 634–642
23. Courteille F, Durou J-D, Morin G (2006) A global solution to the SFS problem using B-spline surface and simulated annealing. In: Proceedings of the international conference on pattern recognition (volume II), pp 332–335
24. Crandall MG, Lions P-L (1983) Viscosity solutions of Hamilton-Jacobi equations. Trans Am Math Soc 277(1):1–42
25. Cristiani E (2014) 3D printers: a new challenge for mathematical modeling. arXiv:1409.1714
26. Cristiani E, Falcone M (2007) Fast semi-Lagrangian schemes for the eikonal equation and applications. SIAM J Numer Anal 45(5):1979–2011
27. Crouzil A, Descombes X, Durou J-D (2003) A multiresolution approach for shape from shading coupling deterministic and stochastic optimization. IEEE Trans Pattern Anal Mach Intell 25(11):1416–1421
28. Cui Z, Gu J, Shi B, Tan P, Kautz J (2017) Polarimetric multi-view stereo. In: Proceedings of the IEEE conference on computer vision and pattern recognition, pp 1558–1567
29. Daniel P, Durou J-D (2000) From deterministic to stochastic methods for shape from shading. In: Proceedings of the Asian conference on computer vision, pp 187–192
30. Durou J-D, Maître H (1996) On convergence in the methods of Strat and of Smith for shape from shading. Int J Comput Vis 17(3):273–289
31. Durou J-D, Falcone M, Sagona M (2008) Numerical methods for shape-from-shading: a new survey with benchmarks. Comput Vis Image Underst 109(1):22–43
32. Falcone M, Ferretti R (2014) Semi-Lagrangian approximation schemes for linear and Hamilton-Jacobi equations, 1st edn. Society for Industrial and Applied Mathematics, Philadelphia
33. Falcone M, Ferretti R (2016) Numerical methods for Hamilton-Jacobi type equations. In: Handbook of numerical methods for hyperbolic problems. Handbook of numerical analysis, vol 17. Elsevier, Amsterdam, pp 603–626
34. Falcone M, Sagona M (1997) An algorithm for the global solution of the shape-from-shading model. In: International conference on image analysis and processing. Lecture notes in computer science, vol 1310, pp 596–603

35. Falcone M, Sagona M (2003) A scheme for the shape-from-shading model with "black shadows". In: Numerical mathematics and advanced applications. Springer, Berlin, pp 503–512
36. Festa A, Falcone M (2014) An approximation scheme for an Eikonal equation with discontinuous coefficient. SIAM J Numer Anal 52(1):236–257
37. Frankot RT, Chellappa R (1988) A method for enforcing integrability in shape from shading algorithms. IEEE Trans Pattern Anal Mach Intell 10(4):439–451
38. Gallardo M, Collins T, Bartoli A (2017) Dense non-rigid structure-from-motion and shading with unknown albedos. In: Proceedings of the IEEE international conference on computer vision, pp 3884–3892
39. Geng J (2011) Structured-light 3D surface imaging: a tutorial. Adv Opt Photonics 3(2):128–160
40. Gotardo PFU, Simon T, Sheikh Y, Matthews I (2015) Photogeometric scene flow for high-detail dynamic 3D reconstruction. In: Proceedings of the IEEE international conference on computer vision, pp 846–854
41. Haefner B, Quéau Y, Möllenhoff T, Cremers D (2018) Fight ill-posedness with ill-posedness: single-shot variational depth super-resolution from shading. In: Proceedings of the IEEE conference on computer vision and pattern recognition, pp 164–174
42. Haefner B, Ye Z, Gao M, Wu T, Quéau Y, Cremers D (2019) Variational uncalibrated photometric stereo under general lighting. In: Proceedings of the IEEE international conference on computer vision, pp 8539–8548
43. Hartley RI, Zisserman A (2004) Multiple view geometry in computer vision, 2nd edn. Cambridge University Press, Cambridge
44. Hernández C (2004) Stereo and silhouette fusion for 3D object modeling from uncalibrated images under circular motion. Thèse de doctorat, École Nationale Supérieure des Télécommunications
45. Herrera SEM, Malti A, Morel O, Bartoli A (2013) Shape-from-polarization in laparoscopy. In: Proceedings of the international symposium on biomedical imaging, pp 1412–1415
46. Hertzmann A, Seitz SM (2005) Example-based photometric stereo: shape reconstruction with general, varying BRDFs. IEEE Trans Pattern Anal Mach Intell 27(8):1254–1264
47. Hold-Geoffroy Y, Gotardo P, Lalonde J-F (2019) Single day outdoor photometric stereo. IEEE Trans Pattern Anal Mach Intell. https://doi.org/10.1109/TPAMI.2019.2962693
48. Horn BKP (1970) Shape from shading: a method for obtaining the shape of a smooth opaque object from one view. PhD thesis, MIT
49. Horn BKP, Brooks MJ (1986) The variational approach to shape from shading. Comput Vis Graph Image Process 33(2):174–208
50. Huynh CP, Robles-Kelly A, Hancock ER (2010) Shape and refractive index recovery from single-view polarisation images. In: Proceedings of the IEEE conference on computer vision and pattern recognition, pp 1229–1236
51. Huynh CP, Robles-Kelly A, Hancock ER (2013) Shape and refractive index from single-view spectro-polarimetric images. Int J Comput Vis 101(1):64–94
52. Ikehata S, Wipf D, Matsushita Y, Aizawa K (2012) Robust photometric stereo using sparse regression. In: Proceedings of the IEEE conference on computer vision and pattern recognition, pp 318–325
53. Ikeuchi K, Horn BKP (1981) Numerical shape from shading and occluding boundaries. Artif Intell 17(1–3):141–184
54. Jancosek M, Pajdla T (2011) Multi-view reconstruction preserving weakly-supported surfaces. In: Proceedings of the IEEE conference on computer vision and pattern recognition, pp 3121–3128
55. Johnson MK, Adelson EH (2011) Shape estimation in natural illumination. In: Proceedings of the IEEE conference on computer vision and pattern recognition, pp 2553–2560
56. Ju Y-C, Tozza S, Breuß M, Bruhn A, Kleefeld A (2013) Generalised perspective shape from shading with Oren-Nayar reflectance. In: Proceedings of the British machine vision conference, pp 42.1–42.11

57. Kadambi A, Taamazyan V, Shi B, Raskar R (2015) Polarized 3D: high-quality depth sensing with polarization cues. In: Proceedings of the IEEE international conference on computer vision, pp 3370–3378
58. Kadambi A, Taamazyan V, Shi B, Raskar R (2017) Depth sensing using geometrically constrained polarization normals. Int J Comput Vis 125:34–51
59. Kong N, Tai Y-W, Shin JS (2013) A physically-based approach to reflection separation: from physical modeling to constrained optimization. IEEE Trans Pattern Anal Mach Intell 36(2):209–221
60. Kontsevich LL, Petrov AP, Vergelskaya IS (1994) Reconstruction of shape from shading in color images. J Opt Soc Am A 11(3):1047–1052
61. Langguth F, Sunkavalli K, Hadap S, Goesele M (2016) Shading-aware multi-view stereo. In: Proceedings of the European conference on computer vision, pp 469–485
62. Leclerc YG, Bobick AF (1991) The direct computation of height from shading. In: Proceedings of the IEEE conference on computer vision and pattern recognition, pp 552–558
63. Li M, Zhou Z, Wu Z, Shi B, Diao C, Tan P (2020) Multi-view photometric stereo: a robust solution and benchmark dataset for spatially varying isotropic materials. IEEE Trans Image Process pp 4159–4173
64. Lions P-L, Rouy E, Tourin A (1993) Shape-from-shading, viscosity solutions and edges. Numer Math 64(1):323–353
65. Logothetis F, Mecca R, Sgallari F, Cipolla R (2019) A differential approach to shape from polarisation: a level-set characterisation. Int J Comput Vis 127(11–12):1680–1693
66. Lyu Y, Cui Z, Li S, Pollefeys M, Shi B (2019) Reflection separation using a pair of unpolarized and polarized images. In: Advances in Neural Information Processing Systems 32, Curran Associates, Inc., pp 14559–14569
67. Mahmoud AH, El-Melegy MT, Farag AA (2012) Direct method for shape recovery from polarization and shading. In: Proceedings of the IEEE international conference on image processing, pp 1769–1772
68. Maurer D, Ju Y-C, Breuß M, Bruhn A (2018) Combining shape from shading and stereo: a joint variational method for estimating depth, illumination and albedo. Int J Comput Vis 126(12):1342–1366
69. Mecca R, Falcone M (2013) Uniqueness and approximation of a photometric shape-from-shading model. SIAM J Imaging Sci 6(1):616–659
70. Mecca R, Wetzler A, Bruckstein A, Kimmel R (2014) Near field photometric stereo with point light sources. SIAM J Imaging Sci 7(4):2732–2770
71. Mecca R, Quéau Y, Logothetis F, Cipolla R (2016) A single-lobe photometric stereo approach for heterogeneous material. SIAM J Imaging Sci 9(4):1858–1888
72. Mélou J, Quéau Y, Castan F, Durou J-D (2019) A splitting-based algorithm for multi-view stereopsis of textureless objects. In: Proceedings of the international conference on scale space and variational methods in computer vision, pp 51–63
73. Miyazaki D, Kagesawa M, Ikeuchi K (2004) Transparent surface modeling from a pair of polarization images. IEEE Trans Pattern Anal Mach Intell 26(1):73–82
74. Miyazaki D, Saito M, Sato Y, Ikeuchi K (2002) Determining surface orientations of transparent objects based on polarization degrees in visible and infrared wavelengths. J Opt Soc Am A 19(4):687–694
75. Miyazaki D, Shigetomi T, Baba M, Furukawa R, Hiura S, Asada N (2012) Polarization-based surface normal estimation of black specular objects from multiple viewpoints. In: Proceedings of the international conference on 3D imaging, modeling, processing, visualization and transmission, pp 104–111
76. Miyazaki D, Shigetomi T, Baba M, Furukawa R, Hiura S, Asada N (2016) Surface normal estimation of black specular objects from multiview polarization images. Opt Eng 56(4):041303
77. Miyazaki D, Tan RT, Hara K, Ikeuchi K (2003) Polarization-based inverse rendering from a single view. In: Proceedings of the IEEE international conference on computer vision, pp 982–987

78. Moons T, Van Gool L, Vergauwen M (2008) 3D reconstruction from multiple images, part 1: principles. Found Trends Comput Graph Vis 4(4):287–404
79. Morel O, Meriaudeau F, Stolz C, Gorria P (2005) Polarization imaging applied to 3D reconstruction of specular metallic surfaces. In: Machine vision applications in industrial inspection XIII. Proceedings of the SPIE, vol 5679, pp 178–186
80. Nayar S, Fang X, Boult T (1997) Separation of reflection components using color and polarization. Int J Comput Vis 21(3):163–186
81. Nayar SK, Nakagawa Y (1994) Shape from focus. IEEE Trans Pattern Anal Mach Intell 16(8):824–831
82. Nehab D, Rusinkiewicz S, Davis J, Ramamoorthi R (2005) Efficiently combining positions and normals for precise 3D geometry. ACM Trans Graph 24(3):536–543
83. Ngo TT, Nagahara H, Taniguchi R (2015) Shape and light directions from shading and polarization. In: Proceedings of the IEEE conference on computer vision and pattern recognition, pp 2310–2318
84. Onn R, Bruckstein AM (1990) Integrability disambiguates surface recovery in two-image photometric stereo. Int J Comput Vis 5(1):105–113
85. Or-El R, Rosman G, Wetzler A, Kimmel R, Bruckstein A (2015) RGBD-fusion: real-time high precision depth recovery. In: Proceedings of the IEEE conference on computer vision and pattern recognition, pp 5407–5416
86. Osher S, Fedkiw R (2003) Level set methods and dynamic implicit surfaces. In: Applied mathematical sciences, vol 153. Springer, Berlin
87. Peng S, Haefner B, Quéau Y, Cremers D (2017) Depth super-resolution meets uncalibrated photometric stereo. In: Proceedings of the IEEE international conference on computer vision workshops, pp 2961–2968
88. Pintus R, Dulecha TG, Ciortan I, Gobbetti E, Giachetti A (2019) State-of-the-art in multi-light image collections for surface visualization and analysis. Comput Graph Forum 38(3):909–934
89. Prados E, Faugeras O (2005) Shape from shading: a well-posed problem? Proceedings of the IEEE conference on computer vision and pattern recognition 2:870–877
90. Quéau Y, Mecca R, Durou J-D (2016) Unbiased photometric stereo for colored surfaces: a variational approach. In: Proceedings of the IEEE conference on computer vision and pattern recognition, pp 4359–4368
91. Quéau Y, Mecca R, Durou J-D, Descombes X (2017) Photometric stereo with only two images: a theoretical study and numerical resolution. Image Vis Comput 57:175–191
92. Quéau Y, Mélou J, Castan F, Cremers D, Durou J-D (2017) A variational approach to shape-from-shading under natural illumination. In: Proceedings of the international workshop on energy minimization methods in computer vision and pattern recognition, pp 342–357
93. Quéau Y, Wu T, Lauze F, Durou J-D, Cremers D (2017) A non-convex variational approach to photometric stereo under inaccurate lighting. In: Proceedings of the IEEE conference on computer vision and pattern recognition, pp 99–108
94. Quéau Y, Durix B, Wu T, Cremers D, Lauze F, Durou J-D (2018) LED-based photometric stereo: modeling, calibration and numerical solution. J Math Imaging Vis 60(3):313–340
95. Quéau Y, Durou J-D, Aujol J-F (2018) Normal integration: a survey. J Math Imaging Vis 60(4):576–593
96. Rahmann S, Canterakis N (2001) Reconstruction of specular surfaces using polarization imaging. In: Proceedings of the IEEE conference on computer vision and pattern recognition, vol 1
97. Robles-Kelly A, Huynh CP (2013) Imaging spectroscopy for scene analysis. Springer, Berlin
98. Santo H, Samejima M, Sugano Y, Shi B, Matsushita Y (2017) Deep photometric stereo network. In: Proceedings of the IEEE international conference on computer vision workshops, pp 501–509
99. Schechner YY (2011) Inversion by P^4: polarization-picture post-processing. Philos Trans R Soc B: Biol Sci 366(1565):638–648
100. Schechner YY (2015) Self-calibrating imaging polarimetry. In: Proceedings of the IEEE international conference on computational photography

101. Schechner YY, Shamir J, Kiryati N (2000) Polarization and statistical analysis of scenes containing a semireflector. J Opt Soc Am A 17(2):276–284
102. Sengupta S, Kanazawa A, Castillo CD, Jacobs DW (2018) SfSNet: learning shape, reflectance and illuminance of faces in the wild. In: Proceedings of the IEEE conference on computer vision and pattern recognition, pp 6296–6305
103. Sethian JA (1999) Level set methods and fast marching methods. Cambridge monographs on applied and computational mathematics, vol 3, 2nd edn. Cambridge University Press, Cambridge
104. Shabayek AER, Demonceaux C, Morel O, Fofi D (2012) Vision based UAV attitude estimation: progress and insights. J Intell Robot Syst 65(1–4):295–308
105. Shafer SA, Kanade T (1983) Using shadows in finding surface orientations. Comput Vis Graph Image Process 22(1):145–176
106. Shi B, Tan P, Matsushita Y, Ikeuchi K (2013) Bi-polynomial modeling of low-frequency reflectances. IEEE Trans Pattern Anal Mach Intell 36(6):1078–1091
107. Shi B, Mo Z, Wu Z, Duan D, Yeung S, Tan P (2019) A benchmark dataset and evaluation for non-Lambertian and uncalibrated photometric stereo. IEEE Trans Pattern Anal Mach Intell 41(2):271–284
108. Shurcliff WA (1962) Polarized light, production and use. Harvard University Press, Harvard
109. Simchony T, Chellappa R, Shao M (1990) Direct analytical methods for solving Poisson equations in computer vision problems. IEEE Trans Pattern Anal Mach Intell 12(5):435–446
110. Smith WAP, Fang F (2016) Height from photometric ratio with model-based light source selection. Comput Vis Image Underst 145:128–138
111. Smith WAP, Ramamoorthi R, Tozza S (2016) Linear depth estimation from an uncalibrated, monocular polarisation image. In: European conference on computer vision. Lecture notes in computer science, vol 9912, pp 109–125
112. Smith WAP, Ramamoorthi R, Tozza S (2019) Height-from-polarisation with unknown lighting or albedo. IEEE Trans Pattern Anal Mach Intell 41(12):2875–2888
113. Soukup D, Huber-Mörk R (2014) Convolutional neural networks for steel surface defect detection from photometric stereo images. In: International symposium on visual computing. Lecture notes in computer science, vol 8887, pp 668–677
114. Sun J, Smith M, Smith L, Coutts L, Dabis R, Harland C, Bamber J (2008) Reflectance of human skin using colour photometric stereo: with particular application to pigmented lesion analysis. Skin Res Technol 14(2):173–179
115. Szeliski R (1991) Fast shape from shading. Comput Vis Graph Image Process: Image Underst 53(2):129–153
116. Taamazyan V, Kadambi A, Raskar R (2016) Shape from mixed polarization. arXiv:1605.02066
117. Tankus A, Sochen N, Yeshurun Y (2003) A new perspective [on] shape-from-shading. In: Proceedings of the IEEE international conference on computer vision 2:862–869
118. Tozza S, Falcone M (2016) Analysis and approximation of some shape-from-shading models for non-Lambertian surfaces. J Math Imaging Vis 55(2):153–178
119. Tozza S, Mecca R, Duocastella M, Del Bue A (2016) Direct differential photometric stereo shape recovery of diffuse and specular surfaces. J Math Imaging Vis 56(1):57–76
120. Tozza S, Smith WAP, Zhu D, Ramamoorthi R, Hancock ER (2017) Linear differential constraints for photo-polarimetric height estimation. In: Proceedings of the IEEE international conference on computer vision, pp 2298–2306
121. Tuchin VV, Wang L, Zimnyakov DA (2006) Optical polarization in biomedical applications. Springer Science & Business Media, New York
122. Vogiatzis G, Hernández C, Cipolla R (2006) Reconstruction in the round using photometric normals and silhouettes. In: Proceedings of the IEEE conference on computer vision and pattern recognition 2:1847–1854
123. White R, Forsyth D (2006) Combining cues: shape from shading and texture. In: Proceedings of the IEEE conference on computer vision and pattern recognition 2:1809–1816

124. Wieschollek P, Gallo O, Gu J, Kautz J (2018) Separating reflection and transmission images in the wild. In: Proceedings of the European conference on computer vision, pp 89–104
125. Witkin AP (1981) Recovering surface shape and orientation from texture. Artif Intell 17(1–3):17–45
126. Wolff LB (1990) Surface orientation from two camera stereo with polarizers. In: Optics, illumination, and image sensing for machine vision IV. Proceedings of the SPIE, vol 1194, pp 287–298
127. Wolff LB (1997) Polarization vision: a new sensory approach to image understanding. Image Vis Comput 15(2):81–93
128. Wolff LB, Boult TE (1991) Constraining object features using a polarization reflectance model. IEEE Trans Pattern Anal Mach Intell 13(7):635–657
129. Woodham RJ (1980) Photometric method for determining surface orientation from multiple images. Opt Eng 19(1):134–144
130. Worthington PL, Hancock ER (1999) Needle map recovery using robust regularizers. Image Vis Comput 17(8):545–557
131. Wu L, Ganesh A, Shi B, Matsushita Y, Wang Y, Ma Y (2010) Robust photometric stereo via low-rank matrix completion and recovery. In: Proceedings of the Asian conference on computer vision, pp 703–717
132. Yang L, Tan F, Li A, Cui Z, Furukawa Y, Tan P (2018) Polarimetric dense monocular SLAM. In: Proceedings of the IEEE conference on computer vision and pattern recognition, pp 3857–3866
133. Yuille AL, Snow D, Epstein R, Belhumeur PN (1999) Determining generative models of objects under varying illumination: shape and albedo from multiple images using SVD and integrability. Int J Comput Vis 35(3):203–222
134. Zhang R, Tsai P-S, Cryer JE, Shah M (1999) Shape-from-shading: a survey. IEEE Trans Pattern Anal Mach Intell 21(8):690–706
135. Zhu D, Smith WAP (2019) Depth from a polarisation + RGB stereo pair. In: Proceedings of the IEEE conference on computer vision and pattern recognition, pp 7586–7595

Chapter 2
Perspective Shape from Shading

An Exposition on Recent Works with New Experiments

Michael Breuß and Ashkan Mansouri Yarahmadi

Abstract Shape from Shading (SFS) is a fundamental task in computer vision. By given information about the reflectance of an object's surface and the position of the light source, the SFS problem is to reconstruct the 3D depth of the object from a single grayscale 2D input image. A modern class of SFS models relies on the property that the camera performs a perspective projection. The corresponding perspective SFS methods have been the subject of many investigations within the last years. The goal of this chapter is to give an overview of these developments. In our discussion, we focus on important model aspects, and we investigate some prominent algorithms appearing in the literature in more detail than it was done in previous works.

Keywords Shape from Shading · Perspective projection · Hamilton Jacobi equations · Numerical methods · Fast marching method

2.1 Introduction

A fundamental task in computer vision with many important applications is to compute the three-dimensional (3D) shape of one or more objects depicted in a photographed scene. Since an image is a projected 2D representation of the 3D world, the information on the 3D geometry is lost during the image acquisition process. Therefore, when given only one input image, the shape reconstruction task amounts to solving a difficult inverse problem.

Shape from Shading (SFS). Shape from Shading is a classic example of techniques for monocular 3D reconstruction. Given a single grayscale input image, the process relies on the *shading*, i.e. on the variation of gray values that appears when light is reflected

M. Breuß · A. Mansouri Yarahmadi (✉)
Applied Mathematics Group, Institute for Mathematics, BTU Cottbus-Senftenberg, Platz der Deutschen Einheit 1, HG 2.51, 03046 Cottbus, Germany
e-mail: Yarahmadi@b-tu.de

M. Breuß
e-mail: Breuss@b-tu.de

© Springer Nature Switzerland AG 2020
J.-D. Durou et al. (eds.), *Advances in Photometric 3D-Reconstruction*,
Advances in Computer Vision and Pattern Recognition,
https://doi.org/10.1007/978-3-030-51866-0_2

at a smooth object surface with uniform reflectance properties. By the latter, SFS may only be applied in the context of non-textured objects since the presence of texture is equivalent to variable reflectance. Therefore, SFS is conceptually orthogonal to Shape from Texture [16, 17, 19, 45] which is also a single-view method.

Since shading is related to the perceived brightness, it is called a photometric cue. The first attempts to gain information on the 3D shape of a surface by such a cue dealt with an astronomical application, namely the reconstruction of the surface of the moon [46, 57]. Then in the PhD thesis of Horn [24]; the problem was investigated systematically and the name SFS was coined. Moreover, SFS was formulated for the first time in terms of a *partial differential equation (PDE)*. The classic PDE-based SFS model of Horn and related techniques have been the subject of extensive investigation during the last decades, see, e.g. [13, 23, 26, 67] for discussions.

Besides being a key problem of theoretical interest in computer vision, SFS has many possible applications. Among them are classical fields such as astronomy [46, 59] or terrain reconstruction [5], and they range more recently from medical applications as, for instance, in dentistry [1, 64] or endoscopy [34, 54, 61] over the digitization of documents [9, 10] in order to enable digital content access to other computer vision tasks like face recognition [51] or facial gender classification [62]. As an example for the use of SFS as a building block for computer vision methods, let us mention photometric stereo [60] where multiple input images taken under different lighting conditions are combined by deriving from corresponding SFS equations a system of equations. Furthermore, SFS may be used with a benefit in combination with different computer vision techniques, see, e.g. [33] for a recent approach fusing SFS with stereo vision in a variational framework.

Modeling of Shape from Shading. Since much information is discarded when capturing the 3D world in a grayscale 2D image, the gap in information is filled by imposing modeling assumptions. Standard assumptions are concerned with *(i)* the illumination in the photographed scene, *(ii)* the light reflectance surface properties and *(iii)* the projection that is performed by the camera. In the context of this work, some aspects of the illumination are important, and we will stick to the relatively modern assumption of a perspective camera projection. These components will be combined here with classic Lambertian surface reflectance.

Let us first comment on the reflectance aspect of the modeling process in more detail. The Lambertian surface model goes back to the work *Photometria* by Johann Heinrich Lambert in 1760 [29]. It is the most simple model for surface reflectance and corresponds to the appearance of a very matte surface without shiny highlights. Because of its simplicity, it is a useful tool for the derivation of mathematical models for SFS, and thus it is used in the classic PDE-based model of Horn as well as in many modern approaches. One may also argue that a wide range of surfaces are in practice fairly close to being Lambertian [14]. However, following the argumentation that the model is too simple for some applications [44], also non-Lambertian surface reflection approaches for SFS have been proposed and analyzed in the literature, see, e.g. [2, 4, 8, 31, 55, 58].

Let us turn to the illumination model. In the classic approach of Horn and many of his successors, it is assumed that the light falls from one and the same direction onto the whole scene. This means that the *light vectors* that point from any location on an object's surface to the light source are all parallel. The corresponding idealization is that there is a point light source at infinity which is, e.g. an adequate model for the light from the sun in an application concerned with reconstructing lunar surfaces, cf. [59]. In this work, we mainly consider this setting but we will also briefly discuss the setup employed, among others, by Prados and his co-authors [39–43] in which the light source is located at the optical center of the camera. This may be considered as an idealization of taking photographs using a camera with photoflash.

As indicated, a key issue in the context of this chapter is the camera projection model. As summarized, for instance, in [23, 67], the classic assumption is that the camera performs an orthographic projection, i.e. an orthogonal projection of the third spatial dimension onto the image plane. This is completely adequate if the photographed object is far away from the camera as it is, for instance, the case in the first applications of SFS concerned with astronomical images. However, if the object of interest is relatively close to the camera as may be the case, e.g. in many applications in optical quality assessment, perspective effects may grow to be important. In such a situation, the use of an orthographic projection will cause significant systematic errors as shown, e.g. in [52]; see also the earlier discussion in [3]. In order to tackle this issue, the orthographic camera model should be substituted employing instead a *perspective projection*.

Contents of This Chapter. In this paper, we give an overview of the classic and recent literature concerned with the use of a perspective projection for SFS. By elaborating on these *perspective SFS (PSFS)* methods, we also highlight some important properties of corresponding models. Let us emphasize that this part of our work is supposed to be potentially useful as an introduction into this special SFS topic so that the exposition is relatively detailed.

Furthermore, we discuss in depth two important variants of PSFS that are often cited in the corresponding literature but have to our best knowledge not been compared rigorously, namely the methods of Tankus et al. [54] and Yuen et al. [66]. These methods are of importance as they represent two technically different approaches to a classic PSFS model setting—employing Lambertian surface reflectance together with parallel lighting—that nevertheless relies on the same numerical resolution principle, namely the fast marching method. By performing the comparison of these methods, we close a systematic gap in the literature. Let us note that this gap may have arisen as especially the implementation of the method of Yuen and his co-workers is not trivial. A technical contribution of this work is that we give a tractable account of a working algorithm for the latter method that can be found in the appendix. Furthermore, we show how to apply Lambertian PSFS methods as discussed here for real-world images by proposing a working pipeline that takes into account typical difficulties of such input images. Since many interesting SFS models are still concerned with the Lambertian setting, we think that this exposition helps to foster their applicability.

2.2 On the History of Perspective Shape from Shading

The first evidence of the use of a perspective projection for SFS is to our best knowledge given in the classic work [25], however, it is of a theoretical nature. Horn describes a perspective projection model together with a nearby light source and an arbitrary reflectivity function. In this setup, a general image brightness equation is derived which is solved under more specific assumptions—notably including a simplification to orthographic projection—by the characteristic strip method using the equivalent system of ordinary differential equations (ODEs).

While the perspective projection is common in geometric approaches in computer vision [20] that rely on multiple views on a scene, it has re-entered the SFS literature relatively late.

Penna [37, 38] proposed a local PSFS method for specific surfaces, i.e. a smooth object and an opaque polyhedron, respectively. He considers systems of algebraic equations that describe the local surface geometry and solves them by an iterative minimization approach.

Lee and Kuo [30] formulate an image irradiance equation for PSFS with Lambertian surfaces and a nearby light source. In order to simplify the proceeding, they consider the resulting image formation over triangular surface patches, and for computations they also employ a linear approximation of the Lambertian reflectance map based on the perspective projection. The resulting algorithm is a relatively complex, iterative variational method minimizing the sum of the squared brightness error.

Okatani and Deguchi [34] employ a perspective projection for SFS for endoscopic images. They do not derive a PDE in terms of an image irradiance equation but introduce an evolution equation for equal-range contours so that they can apply an extension of the Kimmel–Bruckstein algorithm [28].

Cho et al. [9] employ a perspective projection in SFS, but in the specific framework employed in the context of their application concerned with document digitization; they assume that distance variations between camera and surface can be ignored, separating perspective and shading effects.

Hasegawa and Tozzi [21] as well as Yamany et al. [63, 64] consider the simultaneous procedure of camera calibration and PSFS in multi-image settings which results in specific, iterative frameworks. Let us note that these approaches rely on linearizations of the SFS model since, otherwise, computations are too cumbersome and the complete models would be highly complex. Also, SFS for one input image alone is not sufficient to provide both the depth and the camera parameters. Therefore, some additional technical step is required as, for example, realized by a learning algorithm using neural networks [64].

Samaras and Metaxas [47] consider a deformable shape model that results in a highly sophisticated framework for SFS together with illumination direction estimation which allows incorporating a perspective projection.

Based on the general, well-known format of the Lambertian image irradiance equation, Yuen et al. [65, 66] describe an efficient algorithm for PSFS. In contrast to later works on PSFS, they derive the formula of their algorithm in a completely discrete setting by employing simple finite difference expressions based on point coordinates under perspective projection. A more detailed version of this work is given in [66]. We will give a detailed account of this method later.

Nearly simultaneous and independent from each other, three different groups introduced general PSFS models formulated explicitly by PDEs equivalent to the corresponding image irradiance equation [11, 12, 41, 53]. These relatively influential models in the field have as the main ingredients the perspective camera projection, Lambertian surface reflectance and parallel lighting from infinity. In our exposition, we will focus on the method of Tankus and co-authors [53, 54] as a representative of the PDE-based approach.

As a further important step in the development of PSFS, Prados and Faugeras considered a point light source in the finite range of the photographed scene and introduced a *light attenuation factor* that takes the form of an inverse square law in the distance of a surface point to the point light source position. This enabled some degree of well-posedness as discussed in [6, 39, 40]. We also briefly recall this model here.

In the context of some of the mentioned modern works, interestingly already Horn noted in [25] that both the constant lighting direction as well as an inverse square law are possible variations with his perspective approach. Thus the ideational origin of corresponding elements of these works should be credited in the conceptual literature more clearly to Horn.

2.3 Derivation of the Perspective Image Irradiance Equation

In this section, we recall the model for PSFS as introduced in [24, 34] and the constituting *image irradiance equation* as derived—in various formats—in [11, 12, 41, 53]. In order to emphasize on an introductory value of the chapter, we perform the derivation in some detail.

2.3.1 Mathematical Setup

Let $(x, y) \in \mathbb{R}^2$ be in the image domain Ω which is part of the *retinal plane*, where Ω is an open set. Furthermore,

- $u = u(x, y)$ denotes the unknown, sought depth map, as specified below.
- $I = I(x, y)$ is the normalized brightness function. It is sometimes written as $I = \frac{E(x,y)}{\sigma}$, where $E \in [0, 255]$ is the graylevel of the given image and σ is the product of the surface albedo—i.e. the degree by which the surface reflects light—and the light source intensity. Since many physical assumptions on lighting and simplifications

with respect to. the lens system of the camera are underlying the complete process, cf. [26], I is usually normalized to be in $[0, 1]$.

Let us note that when considering the whole image formation process, it is sometimes recommended to distinguish carefully between the irradiance, i.e. the amount of light falling on a surface, and the radiance, i.e. the amount of light radiated from a surface. Because it is possible to argue that these quantities are proportional and since a normalization is involved as discussed, in practice, we usually identify these quantities and consider I as the normalized input image. In other words, I contains the given data. See, e.g. [26] for a detailed discussion of the fundamentals.

– f is the focal length, i.e. the distance between the *optical center C* of the camera and the 2D plane to which the scene of interest is mapped. In SFS, it is usually assumed as a given parameter; we also assume this here.

Let M be a generic point on an object's surface Σ. The unknown of the problem is the function $u : \Omega \to \mathbb{R}$ such that

$$M = M(x, y) = u(x, y)\, m' \tag{2.1}$$

where

$$m' = \frac{f}{\sqrt{x^2 + y^2 + f^2}}\, m \quad \text{and} \quad m = (x, y, -f)^\top, \tag{2.2}$$

with their geometrical representations shown in Fig. 2.1. Verbalizing these definitions, when considering the position vectors of the points on the ball with radius f around C, then $u(x, y)$ describes pixelwise a factor by which the position vector pointing to $(x, y, -f)$ is stretched in order to give M. Note in this context that $u > 0$ holds as the depicted scene is in front of the camera, and that $u(x, y)$ is measured in terms of multiples of f. Concluding this paragraph, let us note that there is implicitly an important modeling step that has taken place, namely that the origin of the coordinate system is identified with the camera position. This is not self-evident, and also other choices may make sense depending on the experimental setting, see, e.g. [15, 27].

Fig. 2.1 A schematic 2D view of a 3D scene containing a Lambertian surface parametrized by Prados and Faugeras [41]

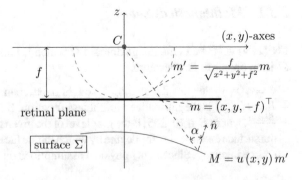

2.3.2   On the Image Irradiance Equation

The PDEs associated with this PSFS model are derived using the *image irradiance equation*, cf. [26]:

$$R(\hat{n}(x, y)) = I(x, y). \tag{2.3}$$

The reflectance function R yields the value of the light reflection on the surface in dependence on the normal $\hat{n}$ of the unknown object surface.

When defining the reflectance function, it is in SFS usually assumed that the illumination is already given. As indicated, we will consider for the systematic discussion two types of illumination, namely parallel lighting as by a point light source at infinity (which will also be the setting of our main experiments) and a point light source nearby the photographed scene. This yields two reflectance functions R_p and R_n depending on the setting. In order to define these functions, we make use of the *light vector ω* and the *distance function r*, respectively.

Taking at this point additionally the assumption of a *Lambertian surface* into account, the function R_p for parallel lighting is defined as

$$R_p(\hat{n}(x, y)) = \omega \cdot \hat{n}(x, y) \tag{2.4}$$

where ω is a constant unit vector pointing from the scene to the light source. Since it is for all points on an object surface the same ω, this gives a bundle of parallel light vectors.

Let us consider now an aspect of the distance function r which is also discussed, e.g. in [34]. If the light source is assumed as above to be distant from the object surface, the distance r between a surface point and the light source turns out to be approximately a constant for any of those points. This situation changes if the light source is assumed to be close to the object surface. In a physically motivated modeling of the latter situation, this implies that an additional factor $1/(4\pi r^2)$ should be taken into account. This term represents the solid angle that corresponds to the unit area on the sphere of radius r centered at the point light source, which means that the incident light reaching an area on the surface decreases by the *inverse square law $1/r^2$*. See, e.g. [26] for a detailed description of the concept of the solid angle.

Furthermore, in the direction of defining the reflectance function R_n to be used with a nearby light source, the light vector ω now depends on x and y since the lighting direction is different for distinct points on a lighted object surface. Taking then again the assumption of a *Lambertian surface* into account, the function R_n for a nearby point light source is defined as

$$R_n(\hat{n}(x, y)) = \frac{\omega(x, y) \cdot \hat{n}(x, y)}{r^2(x, y)}. \tag{2.5}$$

Note that this way of writing down the reflectance function as in (2.4) and (2.5) implies a previous proper normalization as indicated above.

2.3.3 How the Perspective Projection Enters the Model

While we will consider the light vector ω as given, the question we deal with now is how to compute the unknown surface normal vector $\hat{n}$ in (2.4) and (2.5), respectively.

To tackle this question, we recall the mathematical setup and note that we have already parametrized the unknown object surface Σ by Eq. (2.1) in a perspective coordinate system:

$$\Sigma := \Sigma(x, y) = u(x, y)m' = \frac{u(x, y)\mathfrak{f}}{\sqrt{x^2 + y^2 + \mathfrak{f}^2}} \begin{pmatrix} x \\ y \\ -\mathfrak{f} \end{pmatrix}. \qquad (2.6)$$

Employing this perspective parameterization, we consider now the following procedure. Taking directional derivatives in x- and y-directions of Σ, we obtain at any surface point $M(x, y)$, two vectors tangential to Σ. That means, these two vectors Σ_x and Σ_y span the tangential plane at $\Sigma(x, y)$. By means of the cross product $\Sigma_x \times \Sigma_y$, we then obtain a normal vector $\mathbf{n}$ of the surface Σ at the point $M(x, y)$. Normalization of $\mathbf{n}$ gives the sought unit normal vector to the surface.

Employing for abbreviation

$$\zeta := \zeta(x, y) := 1/\sqrt{x^2 + y^2 + \mathfrak{f}^2} \qquad (2.7)$$

for the normalization factor incorporated in (2.6), a simple computation yields

$$\Sigma_x = \mathfrak{f}\zeta \begin{bmatrix} u + xu_x - x^2\zeta^2 u \\ yu_x - xy\zeta^2 u \\ -\mathfrak{f}u_x + \mathfrak{f}x\zeta^2 u \end{bmatrix} \quad \text{and} \quad \Sigma_y = \mathfrak{f}\zeta \begin{bmatrix} xu_y - xy\zeta^2 u \\ yu_y + u - x^2\zeta^2 u \\ -\mathfrak{f}u_y + \mathfrak{f}y\zeta^2 u \end{bmatrix}. \qquad (2.8)$$

Evaluating as indicated, the cross product yields

$$\mathbf{n} = \Sigma_x \times \Sigma_y = \begin{bmatrix} \mathfrak{f}u_x \\ \mathfrak{f}u_y \\ xu_x + yu_y \end{bmatrix} - \mathfrak{f}\zeta^2 \begin{bmatrix} xu \\ yu \\ -\mathfrak{f}u \end{bmatrix}. \qquad (2.9)$$

For further use, we slightly summarize the terms in (2.9) writing equivalently

$$\mathbf{n} = \left(\mathfrak{f}\nabla u - u\mathfrak{f}\zeta^2(x, y)^\top, \nabla u \cdot (x, y)^\top + \mathfrak{f}^2\zeta^2 u \right)^\top. \qquad (2.10)$$

Furthermore, one can compute the Euclidean length of $\mathbf{n}$ as

$$|\mathbf{n}| = \sqrt{\mathfrak{f}^2 |\nabla u|^2 + \left(\nabla u \cdot (x, y)^\top \right)^2 + u^2\mathfrak{f}^2\zeta^2}. \qquad (2.11)$$

At this point we choose not to perform the normalization as required for writing down $\hat{n} = \mathbf{n}/|\mathbf{n}|$ completely and postpone this step to the next two paragraphs, respectively.

2.3.4 Perspective SFS with Parallel Light from Infinity

Let us now elaborate in more detail on the case of a *constant* light vector ω written as

$$\omega := (\mathbf{l}, \gamma)^\top \quad \text{with} \quad \mathbf{l} := (\alpha, \beta)^\top \tag{2.12}$$

where we assume that ω is normalized. Evaluating then the scalar product (2.4) and using (2.3), we obtain the constituting PDE of PSFS for a point light source at infinity:

$$I = \frac{\mathbf{l} \cdot \left(\mathsf{f}\nabla u - u\mathsf{f}\zeta^2(x, y)^\top\right) + \gamma\nabla u \cdot (x, y)^\top + \gamma\mathsf{f}^2\zeta^2 u}{\sqrt{\mathsf{f}^2 |\nabla u|^2 + \left(\nabla u \cdot (x, y)^\top\right)^2 + u^2\mathsf{f}^2\zeta^2}}. \tag{2.13}$$

It is worth mentioning as noted already in [12, 41, 53] that the solution of the PSFS process—as determined by the PDE (2.13) or any of its equivalent representations—provides the shape of an object up to a multiplicative scaling factor. This becomes evident when substituting the unknown u in (2.13) by Cu, C a constant.

The equation in (2.13) represents apart from using a different notation exactly the model as derived in [52, 53] by Tankus et al., in [11] by Courteille et al. and in [41] by Prados and Faugeras, respectively.

As proposed for the corresponding model in [42, 52], we perform the change of variable $v = \ln(u)$ using the property $u > 0$. Dividing by u in both nominator and denominator of Eq. (2.13) and using $\nabla v = (1/u)\nabla u$, we obtain

$$I = \frac{\mathbf{l} \cdot \left(\mathsf{f}\nabla v - \mathsf{f}\zeta^2(x, y)^\top\right) + \gamma\nabla v \cdot (x, y)^\top + \gamma\mathsf{f}^2\zeta^2}{\sqrt{\mathsf{f}^2 |\nabla v|^2 + \left(\nabla v \cdot (x, y)^\top\right)^2 + \mathsf{f}^2\zeta^2}} \tag{2.14}$$

simplifying the PDE.

When referring to the works [12, 41, 53], let us note that there are a few differences in the technical details, so that the PDE above does not appear in exactly the format (2.14) as a result of the derivations there. We consider here writing the setup in the style as employed by Prados and Faugeras, e.g. in [42, 43], there in the context of positioning the light source at the optical center of the camera, since this helps to unify the presentation.

In the work [41], the retinal plane and the unknown surface are parameterized at the positive z-domain in 3D space, not as part of the negative z-domain as here, cf. Fig. 2.1. Furthermore, the unknown u is defined there as a factor stretching the vector $(x, y, \mathsf{f})^\top$ and not its corresponding normalized version m', see (2.2), as here.

Regarding the PSFS model of Tankus et al. [52–54], as in [41], the positive z-domain is used for defining the model, but they make use of the classical pinhole camera model where the retinal plane is located behind the camera. Employing then a different style in deriving their PDE, Tankus et al. rely in a very clean way on the use of perspective projection equations relating a real-world Cartesian coordinate system centered at the camera and coordinates on the retinal plane. However, as perceivable especially in the proof of Theorem 1 in the appendix of [52], finally the same surface parameterization as in [41] is employed.

Courteille et al. [11, 12] make use of the central perspective projection in the style of Tankus et al. but where the retinal plane is between the camera and object surface as in the current work, and they reformulate their PDE so that it closely resembles the eikonal equation. It is the specific contribution of their work to explore benefits of that format for the application of removing geometric defects in scans of documents.

2.3.5 Perspective SFS with Nearby Point Light Source at Optical Center

We now consider PSFS for a nearby point light source, specifying ingredients in (2.5). As indicated before, we will not consider this model again in experiments; our main purpose in the presentation is the attempt to give a complete discussion of the modeling basis and to show the different versions of PDEs that may arise.

The most simple choice in the case considered here is to locate the light source at the optical center of the camera which models the situation of having a camera with a photoflash. This assumption implies that there can be no shadows of photographed objects in an image which cannot be modeled by SFS. Moreover, we will obtain a relatively simple PDE. It is of course possible to put the light source elsewhere but this issue involves further considerations, cf. [15].

If the point light source is given as indicated, one can directly write down the light vector $\omega(x, y)$ for any point on the surface $\Sigma(x, y)$ by

$$\omega(x, y) = \frac{(-x, -y, \mathsf{f})^\top}{\sqrt{x^2 + y^2 + \mathsf{f}^2}} \tag{2.15}$$

since the corresponding vector simply points to the origin. Evaluating the inner product $\hat{n}(x, y) \cdot \omega(x, y)$ using (2.15) together with (2.10) and (2.11) gives after some computations

$$\hat{n}(x, y) \cdot \omega(x, y) = \frac{\mathsf{f}\zeta u(x, y)}{|\mathbf{n}|}. \tag{2.16}$$

Since the distance function r to be used in this setting is given by $r = \mathsf{f}u$, cf. (2.5), we obtain by (2.3)

$$I\sqrt{\mathsf{f}^2 |\nabla u|^2 + \left(\nabla u \cdot (x, y)^\top\right)^2 + u^2 \mathsf{f}^2 \zeta^2} = \frac{\zeta}{\mathsf{f}u(x, y)}. \tag{2.17}$$

As proposed in [42, 52], we perform the change of variable $v = \ln(u)$ using the property $u > 0$. Dividing by u on both sides of Eq. (2.17) and using $\nabla v = (1/u)\nabla u$, we obtain

$$I\sqrt{f^2 |\nabla v|^2 + \left(\nabla v \cdot (x, y)^\top\right)^2 + f^2\zeta^2} = \frac{\zeta}{f}e^{-2v}. \qquad (2.18)$$

This is exactly the PDE as derived by Prados and Faugeras in [43].

Let us note that in a development step documented in [42], Prados and Faugeras proposed the above simplification of putting the light source at the optical center but did not incorporate the light attenuation term $1/r^2$. However, the idea of putting the light source to the image projection center in combination with using the light attenuation term is not new and was proposed by Okatani and Deguchi [34] while they did not write their model in terms of a PDE.

2.3.6 A Discrete PSFS Model Based on the Fast Marching Scheme

Slightly earlier than the models presented above, a related PSFS method has been presented by Yuen, Tsui, Leung and Chen in [66]. Besides the model that we will briefly discuss, an important point in that work is the use of the *fast marching (FM) method*. In order to introduce the scheme of Yuen et al., we briefly elaborate on the FM method first, anticipating in this point the following section on PSFS algorithms.

2.3.6.1 Fast Marching

The idea as it can be understood in the context of SFS is to advance a front from the foreground of the depicted object to the background. Associating thereby each pixel of the input image with a 3D depth value, pixels are distinguished by the labels known, trial and far, respectively. For initialization, all pixels are labeled as far defining their depth values as infinity. This has to be understood as a preliminary value useful for computation when thinking especially of the foreground pixels.

Since the FM method propagates information from the foreground to the background, the scheme relies on depth values being supplied in pixels that are the most in the foreground, i.e. pixels with (locally) minimum depth. The corresponding points are given by the local brightness maxima in the input image and are called *singular points*; see, e.g. the discussion in [35]. Such singular points are marked as trial, which concludes the initialization of the method.

The trial candidate with the smallest computed depth is in the next step marked as known. This involves fixing the depth there for the complete computation since a known pixel is not revisited during the computation. The pixels in terms of the finite difference stencil to the new set of known points are updated with respect to their label, marking them as trial. This process is repeated until all image pixels have the label known. The described procedure involves solving a nonlinear equation in each pixel, for which purpose, e.g. the regula falsi can be employed conveniently.

The concept of the FM method has been brought up by Tsitsiklis in 1995 [56] for the computation of optimal trajectories as solutions of Hamilton–Jacobi equations. The same concept has also been developed by Helmsen et al. [22] and Sethian [48, 50] in 1996. In contrast to Tsitsiklis' method, the latter two schemes were proposed for solving the eikonal equation, and they rely on the idea of using *upwind discretizations*, i.e. one-sided finite differences oriented in accordance with wave propagation direction.

While the three methods have been developed apparently independently of each other, it was Sethian who coined the name *fast marching*. The FM method is computationally very efficient since it is a single-pass method, i.e. as indicated each pixel is visited only once during the computation. Therefore, given n pixels in an image, it has complexity $O(n \log n)$ where the $\log n$ arises since the computed depth values of the trial pixels have to be sorted in each step of the method; a canonical choice for doing this is the heap sort algorithm. As a general reference on FM, let us mention [49] while we refer to [7] for a recent discussion of FM methods.

Let us note that a crucial step of the FM method is to provide a suitable initial depth value at singular points. Since the corresponding procedure differs depending on the model, we choose to discuss that issue in the algorithmic section.

2.3.6.2 Discrete FM-Based Model for PSFS

In this subsection, we review two methods, namely Yuen et al. [66] and Tankus et al. [54], that adopt fast marching [49] for the perspective SFS scenario.

Let us start with a review of the model assumptions used by Yuen et al. [66], and later we refer to the other method.

The sought surface in [66] is assumed to be Lambertian illuminated by a far away light source. A camera is located at the center of a reference Cartesian coordinate system and projects a perspective image of the surface to the retinal plane that is placed at the negative side of the z-axis. The camera has a focal length of f. These sets of assumptions draw an analogous modeling process compared to that considered by Tankus et al. [52–54] which is the main motivation for us to evaluate them against each other.

We proceed to show the derivation of the normal vectors. Yuen et al. [66] take a central surface point $P_0 (x_0, y_0, z_0)$ along with its four neighbors P_1, P_2, P_3 and P_4 (see Fig. 2.2) and their corresponding 2D perspective points projected on a uniform grid

$$P_0 \left(x_0, y_0, z_0\right) = \left(\tfrac{u_0}{f} z_0, \tfrac{v_0}{f} z_0, z_0\right)^T := \left(\tfrac{u}{f} z_{u,v}, \tfrac{v}{f} z_{u,v}, z_{u,v}\right)^T,$$

$$P_1 \left(x_1, y_1, z_1\right) = \left(\tfrac{u_1}{f} z_1, \tfrac{v_1}{f} z_1, z_1\right)^T := \left(\tfrac{u-1}{f} z_{u-1,v}, \tfrac{v}{f} z_{u-1,v}, z_{u-1,v}\right)^T,$$

$$P_2 \left(x_2, y_2, z_2\right) = \left(\tfrac{u_2}{f} z_2, \tfrac{v_2}{f} z_2, z_2\right)^T := \left(\tfrac{u}{f} z_{u,v-1}, \tfrac{v-1}{f} z_{u,v-1}, z_{u,v-1}\right)^T,$$

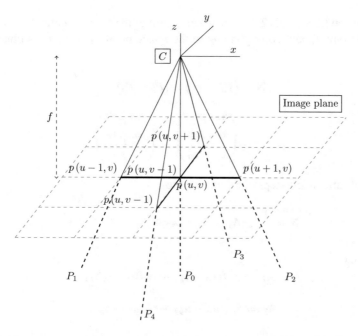

Fig. 2.2 The image coordinate system used by Yuen et al. [66] with its origin at the center of the perspective projection C. The neighbor points P_0, P_1, P_2, P_3 and P_4 located on a Lambertian surface are mapped to their corresponding points on image plane using (2.19)

$$P_3\left(x_3, y_3, z_3\right) = \left(\tfrac{u_3}{f}z_3, \tfrac{v_3}{f}z_3, z_3\right)^{\top} := \left(\tfrac{u+1}{f}z_{u+1,v}, \tfrac{v}{f}z_{u+1,v}, z_{u+1,v}\right)^{\top},$$

$$P_4\left(x_4, y_4, z_4\right) = \left(\tfrac{u_4}{f}z_4, \tfrac{v_4}{f}z_4, z_4\right)^{\top} := \left(\tfrac{u}{f}z_{u,v+1}, \tfrac{v+1}{f}z_{u,v+1}, z_{u,v+1}\right)^{\top} \quad (2.19)$$

and generalize them in the form of

$$P_a := \left(\tfrac{u_a z_a}{f}, \tfrac{v_a z_a}{f}, z_a\right)^{\top} \quad \text{with} \quad a \in \{1, 3\} \quad (2.20)$$

and

$$P_b := \left(\tfrac{u_b z_b}{f}, \tfrac{v_b z_b}{f}, z_b\right)^{\top} \quad \text{with} \quad b \in \{2, 4\}. \quad (2.21)$$

Note that the sought depth in (2.19) is $z_0 := z_{u,v}$, and we make use of the notion of z_0 while discussing the model proposed by Yuen et al. [66] in coming paragraphs. Let us emphasize here that we employ the notation z_0 only in order to simplify the presentation, i.e. in order to avoid a lot of double indices that may arise. The index of z_0 should just indicate that the center of the computational stencil is considered.

Based on (2.20) and (2.21), the generic form of the outward normal vector at the surface point P_0 and corresponding to the image point $p_0(u, v, f)$ is obtained in [66] as

$$\mathbf{N} = (P_a - P_0) \times (P_b - P_0)$$

$$= \begin{pmatrix} \frac{u_a z_a - u_0 z_0}{f} \\ \frac{v_a z_a - v_0 z_0}{f} \\ z_a - z_0 \end{pmatrix} \times \begin{pmatrix} \frac{u_b z_b - u_0 z_0}{f} \\ \frac{v_b z_b - v_0 z_0}{f} \\ z_b - z_0 \end{pmatrix}$$

that leads after a few steps of calculations to

$$\mathbf{N} = \frac{1}{f} \left(z_0 A_1 + B_1, z_0 A_2 + B_2, z_0 A_3 + B_3 \right)^\top \tag{2.22}$$

by having

$$A_1 := z_a \left((v_0 - v_a) - z_b (v_0 - v_b) \right),$$

$$A_2 := z_b \left((u_0 - u_b) - z_a (u_0 - u_a) \right),$$

$$A_3 := \frac{z_b}{f} \left((u_b v_0 - u_0 v_b) + z_a (u_0 v_a - u_a v_0) \right),$$

and

$$B_1 := z_a z_b (v_a - v_b),$$

$$B_2 := z_a z_b (u_b - u_a),$$

$$B_3 := \frac{z_a z_b}{f} (v_b u_a - u_b v_a).$$

To this end, by substituting the derived normal (2.22) and the vertical light source direction $(0, 0, 1)$ into the image irradiance equation (2.4), Yuen et al. [66] derive the image irradiance equation as

$$I_0 = \frac{|z_0 A_3 + B_3|}{\sqrt{(z_0 A_1 + B_1)^2 + (z_0 A_2 + B_2)^2 + (z_0 A_3 + B_3)^2}}. \tag{2.23}$$

Yuen et al. [66] proceed by solving (2.23) in the form of quadratic equations (2.48) and (2.57) from which the unknown depth z_0 is found as explained in Algorithms 1, 2 and 3 and their corresponding Appendices 1–3.

Let us now turn for comparison to the method of Tankus et al. [54] by following the setup shown in Fig. 2.3. Here, a surface S located at the positive side of the z-axis is projected to an image plane uv. The uv plane is located at the distance f from the xy plane and at the negative side of the z-axis. The surface itself is defined

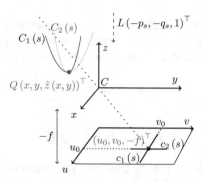

Fig. 2.3 The perspective scenario proposed by Tankus et al. [54], where a surface point $Q\left(x, y, \hat{z}\left(x, y\right)\right)$ is projected to $P\left(u_0, v_0, -f\right)$ on image plane uv. Two curves $C_1\left(u\right)$ and $C_2\left(v\right)$ on surface S passing through Q are projected to their counterparts $c_1\left(s\right)$ and $c_2\left(s\right)$ inside the uv plane. The variable s acts here as a free parameter we use for curve parameterization

as $S = \{\left(x, y, \hat{z}\left(x, y\right)\right)^{\top} \mid \left(x, y\right) \in \Omega_{\text{scene}}\}$, where Ω_{scene} is an open domain. It is further assumed that the surface S is differentiable with respect to both $\left(x, y\right)$ and $\left(u, v\right)$. Based on the perspective projection, one could write the surface as

$$S = \left\{ \left(\frac{-uz}{f}, \frac{-vz}{f}, z\left(u, v\right) \right)^{\top} \mid \left(u, v\right) \in \Omega_{\text{image}} \right\}. \tag{2.24}$$

It is assumed that $\hat{z}\left(x, y\right) = z\left(u, v\right)$. In what follows, Tankus et al. [54] proceed with two major assumptions before starting to derive their model in form of a PDE:

- Surface S is Lambertian and visible from all points of Ω_{image} under the perspective projection.
- A point light source located at infinity illuminates the scene from the direction $\left(-p_s, -q_s, 1\right)^{\top}$.

To start, a parametric curve $c\left(s\right)$ with parameter s is defined over the uv plane

$$c\left(s\right) := \left(u\left(s\right), v\left(s\right), -f\right)^{\top} \tag{2.25}$$

that is considered to be the result of perspective projection of another curve located on surface S

$$C\left(s\right) = \left(\frac{-u\left(s\right)z\left(s\right)}{f}, \frac{-v\left(s\right)z\left(s\right)}{f}, z\left(s\right) \right)^{\top} = \frac{z\left(s\right)}{f}\left(-u, -v, f\right)^{\top} \tag{2.26}$$

whose tangent is derived to be

$$\frac{d}{ds}C\left(s\right) = \frac{1}{f}\left(-u_s\left(s\right)z\left(s\right) - u\left(s\right)z_s\left(s\right), -v_s\left(s\right)z\left(s\right) - v\left(s\right)z_s\left(s\right), f z_s\left(s\right)\right)^{\top}. \tag{2.27}$$

To simplify the scenario, Tankus et al. [54] take a projected point $P = (u_0, v_0, -f)$ on uv plane and define two curves c_1 and c_2 passing through it such that the latter one is parallel to v-axis and the former one to be parallel to u-axis (see Fig. 2.3):

$$c_1(u) = (u, v_0, -f)^\top, \tag{2.28}$$

and

$$c_2(v) = (u_0, v, -f)^\top. \tag{2.29}$$

Referring to Fig. 2.3, let us note that we just specified here the free parameter s as it can be identified in accordance with the axes. One writes the corresponding 3D curves C_1 and C_2 to those 2D counterparts c_1 and c_2 as

$$C_1(u) = \left(\frac{-uz}{f}, \frac{-v_0 z}{f}, z\right)^\top = \frac{z}{f}(-u, -v_0, f)^\top, \tag{2.30}$$

and

$$C_2(v) = \left(\frac{-u_0 z}{f}, \frac{-vz}{f}, z\right)^\top = \frac{z}{f}(-u_0, -v, f)^\top. \tag{2.31}$$

Note again, C_1 and C_2 are now parameterized based on u and v, and not s anymore.

Adopting (2.27), the tangents to both curves (2.30) and (2.31) are written as

$$
\begin{aligned}
\frac{d}{du} C_1(u) &= \frac{d}{du}\left(\frac{1}{f}(-uz, -v_0 z, fz)\right)^\top \\
&= \frac{1}{f}(-z - uz_u, -v_0 z_u, fz_u)^\top,
\end{aligned} \tag{2.32}
$$

and

$$
\begin{aligned}
\frac{d}{dv} C_2(v) &= \frac{d}{dv}\left(\frac{1}{f}(-u_0 z, -vz, fz)\right)^\top \\
&= \frac{1}{f}(-u_0 z_v, -z - vz_v, fz_v)^\top,
\end{aligned} \tag{2.33}
$$

and finally the normal vector $\mathbf{n}$ and its unit form $\hat{\mathbf{n}}$ to the surface S are derived, based on (2.32) and (2.33), as presented in Appendix 4.

Finally, by considering (2.4) and the light source L $(-p_s, -q_s, 1)^\top$, one computes the irradiance image proposed by Tankus et al. [54] as

$$
\begin{aligned}
I &= L \cdot \hat{\mathbf{n}} \\
&= \frac{(-p_s, -q_s, 1)^\top \cdot (fz_u, fz_v, z + vz_v + uz_u)^\top}{\underbrace{\sqrt{p_s^2 + q_s^2 + 1}}_{\|L\|}\sqrt{f^2(z_u^2 + z_v^2) + (z + vz_v + uz_u)^2}}
\end{aligned}
$$

$$= \frac{(u - fp_s)\, z_u + (v - fq_s)\, z_v + z}{\|L\|\sqrt{f^2 \left(z_u^2 + z_v^2\right) + (z + vz_v + uz_u)^2}}. \tag{2.34}$$

Here, $z(u, v) := \hat{z}(x, y)$ with (u, v) is located on the image plane to be the perspective projection of the scene point $(x, y, \hat{z}(x, y))$.

One may argue at this point that the model described in (2.34) may appear as a standard equation in the area of SFS. However, as we aimed to clarify already via some remarks, many models that rely on a similar set of model assumptions differ in the details, for example, by employing specific coordinate systems. The setup as described above now leads in further steps to the approach of Tankus and co-authors, which makes the complete method unique and specific. In particular, as shown by them it is possible to derive a closed-form solution for the slopes of an unknown surface that can be employed for defining an efficient algorithm. The model (2.34) is also in the sense specific that it relies on a PDE, which is not self-evident in the field of SFS where many constructions have been proposed that make not explicit use of a PDE, compare the classic review paper [67].

We proceed by noting that (2.34) shows a direct dependence of I on z, z_u and z_v. By assuming $z(u, v) > 0$ as the depth of the image point with coordinate (u, v), to be always positive, and further replacing the depth $z(u, v)$ with $\ln z(u, v)$, below definitions are provided:

$$\frac{\partial (\ln z(u, v))}{\partial u} = \frac{z_u(u, v)}{z(u, v)} = \frac{z_u}{z} := p, \tag{2.35}$$

and

$$\frac{\partial (\ln z(u, v))}{\partial v} = \frac{z_v(u, v)}{z(u, v)} = \frac{z_v}{z} := q, \tag{2.36}$$

which leads by substituting in (2.34) to

$$I = \frac{(u - fp_s)\, p + (v - fq_s)\, q + 1}{\|L\|\sqrt{f^2 \left(p^2 + q^2\right) + (up + vq + 1)^2}}. \tag{2.37}$$

Note that steps needed to derive (2.37) based on (2.34), (2.35) and (2.36) are further shown in Appendix 5.

In (2.37), one observes that the dependency on $z(u, v)$ is eliminated, and the only dependency is on p and q that are the partial derivatives of $\ln z(u, v)$. Consequently, the problem of obtaining the depth $z(u, v)$ from image irradiance equation (2.34) is now reduced to the problem of deriving $\ln z(u, v)$ from (2.37). As mentioned by [54], the bijective property of the natural logarithm along with the condition $z(u, v) > 0$ makes it possible to recover $z(u, v) = e^{\ln z(u, v)}$.

We proceed with a simplified form of (2.37), derived with a few steps shown in Appendix 6, and which can be written as

$$\alpha_1 + \alpha_2 + \alpha_3 = 0, \tag{2.38}$$

with

$$\alpha_1 := I^2 \|L\|^2 f^2 \left(p^2 + q^2\right) \quad , \quad \alpha_2 := I^2 \|L\|^2 \left(up + vq + 1\right)^2$$

and

$$\alpha_3 := -\left((u - fp_s)\,p + (v - fq_s)\,q + 1\right)^2.$$

To obtain the PDE of Tankus et al. [54], the expanded forms of the α_1, α_2 and α_3 (see Appendix 7) are factorized (see Appendix 8) by their common terms $p^2, q^2, 2pq, 2p$ and $2q$. This leads by rearrangement of (2.38) to

$$p^2 A + q^2 B + 2pq C + 2p D + 2q E + F = 0, \tag{2.39}$$

where p and q, defined as (2.35) and (2.36), are the only unknowns of it. By observing p^2 and q^2 in (2.39) to be always positive values, Tankus et al. [54] separate their corresponding coefficients A and B into

$$A_1 := I^2 \|L\|^2 \left(u^2 + f^2\right) \quad , \quad A_2 := -(u - fp_s)$$

and

$$B_1 := I^2 \|L\|^2 \left(v^2 + f^2\right) \quad , \quad B_2 := -(v - fq_s)^2$$

such that A_1 and B_1 are always non-negative (2.39) so that one may write

$$p^2 A_1 + q^2 B_1 = \hat{F} \quad \text{with} \quad \hat{F} := -p^2 A_2 - q^2 B_2 - 2pq C - 2p D - 2q E - F \tag{2.40}$$

that may be interpreted as a form of an eikonal equation. The algorithmic form of the method of Tankus et al. [54] adopting the perspective fast marching to solve (2.40) can be found in the form of Algorithm 4 with needed supplementary explanations in the form of Appendices 4–8.

2.4 Computational Results

In this section, we compare reconstruction results as created via the algorithms constructed by Tankus et al. [54] and Yuen et al. [66]. The common modeling assumption, namely a far away light source and a sought surface with a Lambertian reflectance property, provides the possibility of this comparison.

In order to illustrate a realistic use of SFS methods, we give here a novel experiment concerned with a real-world input image. The computer vision pipeline that is helpful to be employed in such a setting, and which may also be applied in related circumstances, especially includes a preprocessing step that modifies non-Lambertian reflectance components in an input image in the way that one may safely employ Lambertian methods, as in this chapter.

Our sought surface candidate is that of a Buddha bust having a matte appearance that we capture in a darkened room and by a *PiAEK* endoscopic camera equipped with 8 high-brightness light LEDs integrated in the endoscope. Let us stress that this choice also addresses one of the areas of application of SFS methods, namely related to an endoscopic setting applied on our Buddha bust. The camera has a small focal length of approximately 4 millimeter and suits mostly those models [34, 41–43] that assume the point light source and the camera lens center to be concentric. Our intention to use the lighting fitted to an endoscopic camera may look as not to be the right choice while comparing to the models [54, 66] which have as a basic common assumption a far away light source. To resolve this limitation, we follow the same strategy used in [54] and capture

- only the limited face area of the bust, and while
- the bust distance to the camera is below 5 cm.

The above constraints (i) allow the face area to be illuminated with a very limited range of the light beams emanating from the camera, and (ii) reduce the decay of the illumination strength at different parts of the face area. In this way, the emanating light beams from the camera LEDs show a close resemblance to a far away light source that could have illuminated our bust as parallel light beams coming from infinity. The Buddha bust is shown in Fig. 2.4.

To summarize, if the camera and a connected light source close to the optical center are very close to an object of interest, then the photographed part of an object's surface is very small. Moreover, even the inverse square law that one could employ does not give a meaningful contribution in this setting. In total, such a situation is equally well modeled by employing just parallel lighting and no inverse square law.

Let us emphasize that one cannot expect that the light reflectance at the Buddha bust is correctly described by the Lambertian model. Therefore, in order to employ a Lambertian SFS model for reconstruction, we need to consider a projection that takes the non-Lambertian input image and gives it an adequate Lambertian appearance. Since we know that the Buddha bust shows a matte surface, it is reasonable to think of a non-Lambertian model describing matte surfaces and which could help in constructing a Lambertian version of it. To this end, it appears reasonable that the Lambertian irradiance image I could be approximated based on the irradiance values I_{O-N} obtained from the Oren–Nayar reflectance model [36] as this is an adequate model for matte surfaces. In more general reflectance situations, it may be possible to make use in an analogous way as below, to recompute a Lambertian version of the input image, possibly by learning a more general inverse mapping for the reflectance; see, e.g. [18, 32] for such approaches.

Preprocessing Based on Oren–Nayar Reflectance. The Oren–Nayar model is designed to handle rough and matte objects by modeling their surfaces as an aggregation of many infinitesimally small Lambertian patches called facets. We show a schematic side view of a small area $d\epsilon$ corresponding to a rough surface made from a few facets as Fig. 2.5.

Fig. 2.4 The Buddha bust with a matte surface being captured by an endoscopic camera *PiAEK* equipped with 8 high-brightness light LEDs integrated with the endoscope. The camera has a small focal length of approximately 4 millimeter. The bust was located at a distance less than 5 cm while being captured in an intentionally darkened room. The motivation to capture the bust in a very short distance was to make it reasonable to employ the perspective projection

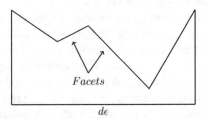

Fig. 2.5 A schematic side view of a small area $d\epsilon$ corresponding to a rough surface aggregating a set of facets. Each facet behaves as a Lambertian surface. The roughness parameter $\sigma \in \left[0, \frac{\pi}{2}\right]$ of the surface is assumed to follow a Gaussian distribution of the whole facet slopes. Based on the Oren and Nayar model, each facet contributes to the irradiance value of the surface as shown in (2.41). Note that, in case of $\sigma = 0$, the surface follows the Lambertian model

The slope values of all such facets comprising the whole rough surface are assumed to follow a Gaussian distribution with a standard deviation $\sigma \in \left[0, \frac{\pi}{2}\right]$ also called *roughness parameter* of the surface. The main idea proposed by Oren and Nayar is that each facet contributes to the modeled irradiance value $I_{\text{O-N}}$ of the surface computed as

$$I_{\text{O-N}} = \frac{\rho}{\pi} L_i \cos(\theta_i)(\nu_1 + \nu_2 \sin(\alpha)\tan(\beta))\max(0, \cos(\Phi_r - \Phi_i)) \qquad (2.41)$$

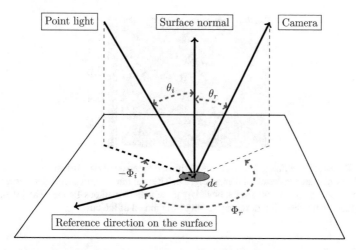

Fig. 2.6 Illustration of the Oren–Nayar model for proposing the reflectance value of a facet being illuminated by a point light source and captured by a camera. The directions in which the facet is captured and illuminated establishes two angles θ_r and θ_i with the normal to the facet, respectively. In addition, the reference direction on the surface establishes two angles Φ_r and Φ_i with the camera and the illumination directions. Note that we could not visualize a particular facet because of its comparatively small size to the small area $d\epsilon$

with the terms ν_1 and ν_2 defined based on the roughness parameter σ as

$$\nu_1 = 1 - 0.5\frac{\sigma^2}{\sigma^2 + 0.33} \quad \text{and} \quad \nu_2 = 0.45\frac{\sigma^2}{\sigma^2 + 0.09}. \qquad (2.42)$$

As one observes in Fig. 2.6, a few further parameters of (2.41) are denoted as ρ to represent the surface albedo, L_i as the intensity of the point light source, θ_i to represent the angle between the surface normal and the light source and θ_r to stand for the angle between the surface normal and the camera direction. In addition, two parameters $\alpha = \max(\theta_i, \theta_r)$ and $\beta = \min(\theta_i, \theta_r)$ represent the maximum and the minimum values of the θ_i and θ_r angles, respectively. Finally, Φ_i and Φ_r denote the angles between the light source and the camera direction each with the reference direction on the surface as shown in Fig. 2.6.

Next we let the point light source to be located at the optical center of the camera and assume the constant coefficient $\frac{\rho}{\pi}L_i$ in (2.41) to be equal to one, because of its dependence only on the light source intensity, surface albedo and the parameters of the imaging system such as the lens diameter and the focal length as mentioned by [2]. This makes us simplify (2.41) to

$$I_{\text{O-N}} = \nu_1 \cos(\theta) + \nu_2 \sin^2(\theta), \qquad (2.43)$$

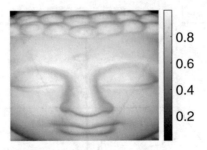

Fig. 2.7 The irradiance image $I \in [0, 1]$ approximated based on (2.44) corresponded to the Buddha bust made from matte materials shown in Fig. 2.4. The maximum irradiance value appears almost on the tip of the nose and is used as the boundary point to both methods Tankus et al. [54] and Yuen et al. [66] and in the direction of reconstructing a 3D model of the bust

while the light source and viewing directions are considered to be pointing in the same direction, resulting in $\theta_i = \theta_r = \theta$ and $\Phi_i = \Phi_r$, with $\cos(\Phi_r - \Phi_i) = 1$ and $\alpha = \beta = \theta$.

Now, one clearly observes that the irradiance values $I_{\text{O-N}}$ of our matte Buddha bust, extracted based on the Oren and Nayar model, consist of two components, namely $\nu_1 \cos(\theta)$ as the Lambertian one and $\nu_2 \sin^2(\theta)$ which is the non-Lambertian component that takes on its maximum value where $\theta = \frac{\pi}{2}$, i.e. close to the occluding boundary.

In proceeding with the actual preprocessing, we follow the work by [44] to solve (2.43) for $\cos(\theta)$ based on the approximated intensity $I_{\text{O-N}}$ as

$$\cos(\theta) = \frac{\nu_1 \pm \sqrt{\nu_1^2 - 4\nu_2(I_{\text{O-N}} - \nu_2)}}{2\nu_2} \tag{2.44}$$

while taking the solutions associated with the minus sign as motivated in [44].

The corrected irradiance image shown as Fig. 2.7 which may be considered as Lambertian is obtained as $I := \nu_1 \cos(\theta)$ from Eq. (2.44). It is used as the input to both models of Tankus and Yuen and respective co-authors while considering the tip of the nose with the highest irradiance value as the boundary point to the perspective fast marching solver tailored by either of the methods [54, 66].

Let us note at this point explicitly that the FM solver is designed to solve boundary value problems of eikonal type. In our application, we have to give an initial depth value which serves as the boundary value. The tip of the nose is the brightest point in the input image and by this, it is a candidate for the smallest depth value as seen from the camera. The FM method then computes subsequently larger depth values starting from the given boundary value. The tip of the nose is identical to a singular point as discussed before in this chapter.

Discussion of Results. The method of Tankus et al. [54] iteratively improves the constructed 3D perspective model of the Buddha face while starting initially from an orthographically created surface of it. We decided to cut the final reconstructed

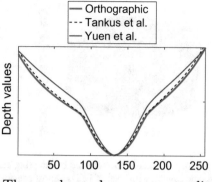

The x values whose corresponding
y values are equal to the y value
of the nose tip

Fig. 2.8 The profile views obtained by cutting the reconstructed 3D surfaces based on Tankus et al. [54] and Yuen et al. [66]. Note that we let the Tankus model iterate 4 times. The cuts are performed horizontally and in the direction of the line passing through the tip of the nose. The point with the minimum depth corresponds to the nose tip. The situation exactly at the nose tip is supposed to be orthographic, so that the method of Tankus et al. shows the correct behavior around it

surface by Tankus in the direction of a horizontal line passing through the tip of the nose and compare it to the same cut taken from the orthographic surface from which the Tankus model starts its evolution. Figure 2.8 shows such horizontal cuts, one taken from the final reconstructed surface after 4 iterations and the other from the orthographic initial state.

Let us stress that now, we first look at a relatively small portion of the image just around the nose tip, so that the presented cut represents a close to orthographic scenario within the experiment. What we expect in just this part of the image is that a correctly working SFS method shall give a solution close to an orthographic method. Any perspective effect may show itself in the final reconstructed profile when moving away from the vicinity of the nose tip, which acts as a common boundary value for all the methods. We observe a correct, close to orthographic behavior of the method of Tankus et al. in Fig. 2.8 as well as some first instances of perspective reconstruction when going away from the nose (but not very strongly). However, the reconstructed profile by Yuen et al. [66], shown also in Fig. 2.8, is observed to have a quite different geometry compared to the other two profiles. This is due to an underlying problem in the method of Yuen et al., which appears when working with a small focal length.

Additionally to the profile views shown in Fig. 2.8, we visualize the reconstructed 3D frontal models of the whole face area with respect to both models of Tankus and Yuen and co-authors as Figs. 2.9 and 2.10, respectively.

As one observes in Fig. 2.9, the point light source of our used endoscopic camera has no negative consequences on the construction of the final model, though Tankus et al. [54] assumes a far away light source and not a point light source in their

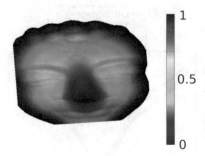

Fig. 2.9 The resultant depth map of Buddha face created by adopting Tankus et al. [54]. Though we have adopted an endoscopic camera with a set of LEDs almost concentric with the camera lens center, the far away light source as one of the main assumptions in Tankus et al. [54] is not being negated. Note that the focal length as one of the parameters in Tankus model is set to the focal length of our adopted camera while inferring the depth

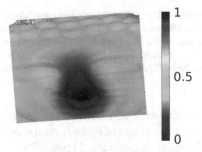

Fig. 2.10 The depth map obtained from the face area of the Buddha bust by adopting Yuen et al. [66]. To infer the depth values, we had to choose the focal length which is one of the model parameters of the method of Yuen et al. [66] as a large number, which means that our endoscopic camera acts as an orthographic one. The proposed model by Yuen et al. [66] may not suit PSFS in endoscopic applications

proposed model. The reason to still have a final smooth surface is the resemblance of the illuminated light by our camera to a typical far away source of light because of locating the Buddha bust at a close vicinity of the camera while capturing it, as already indicated. Let us stress that we used the focal length of our camera while reconstructing the surface shown in Fig. 2.9.

It is noteworthy that, to obtain a complete reconstruction of the Buddha bust based on the proposed method by Yuen et al. [66], which is shown in Fig. 2.10, we had to choose the focal length, which is one of the model parameters, as a large number. This would lead the endoscopic camera model to resemble an orthographic camera that may not fulfill our initial aim to create a perspective model of the Buddha bust. We also aimed to use the real focal length of our camera as the parameter input of the Yuen model, but no model could be created as the final result. The unstable computation reveals the fact that the Yuen model may not be suitable for PSFS while adopting an endoscopic camera with a small focal length and small sensor size.

As indicated by making use of the profile cuts, the perspective effect is supposed to be mainly visible away from the center of a given input image, basically toward the outer parts of a reconstructed geometry. We expect to observe this effect in the now presented synthetic and quantitatively evaluated experiment. By performing this experiment, complementary to the presentation of the profile cuts, we confirm that the method of Tankus et al. is not only in a near-orthographic but also in the truly perspective setting a very robust and trustworthy method from the literature, as proposed of course already in the original papers but here evaluated with the emphasis on the interplay of an orthographic and perspective situation. The latter aspect has been to our impression not investigated in too much detail in the previous literature.

As another experiment, we take the hemisphere shown in Fig. 2.11, also used in [66, 67], as our baseline geometry while being illuminated by the light source at infinity. It is rendered as a Lambertian surface, and we investigate how the approximations obtained by Tankus and Yuen methods deviate from it in terms of l_2 and l_∞ measures. In addition, we aim to vary the focal length of the camera to see which model may be more robust. Starting the detailed discussion and in the direction of making the comparison results simpler to observe, we just cut the baseline geometry and the approximated hemispheres such that the cut passes through the zenith and is vertical to the xy plane.

Now, let us start the experiment with a relatively high focal length of 64, shown as the rightmost graph in the first row of Fig. 2.11, and compare the cut obtained from the baseline geometry shown in red color to the green and the blue cuts taken from the Yuen and the Tankus models. This case looks to be an ideal observation, since both methods have their cuts nearly approximating the baseline cut with a good accuracy. In terms of l_2 and l_∞ measures, we observed the values of 36.97 and 3.82 corresponding to the Tankus model and 13.90 and 0.32 with respect to Yuen's approach.

Next, we reduced the focal length, initially to the value of 32 and later to 16. The corresponding cuts are shown as the middle and the left graph appearing as the first row of Fig. 2.11, respectively. The reduction of focal length clearly indicates the poor performance of the Yuen model in this regime. We observed the same effect while producing the Buddha bust using the Yuen model, that leads us to increase the focal length so the approximation as shown in Fig. 2.10 is produced. In this experiment, the focal length reduction had almost no effect on the model produced by Tankus et al., since the corresponding l_2 and l_∞ measures stay in the vicinity of the already related mentioned values. Note that, in this experiment, we kept the number of iterations taken by the Tankus approach to be equal to 4 again. In the context of the previous experiment with the Buddha bust, let us stress that we observe the validity of the perspective effect away from the image center.

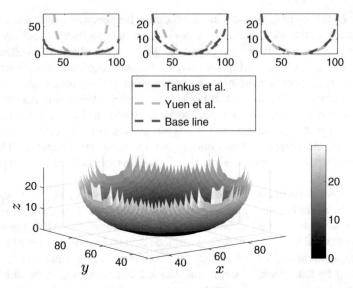

Fig. 2.11 The sought surface of interest, namely a hemisphere, shown as the bottom and used in [66, 67] as the baseline geometry. The rightmost graph in the first row can be considered as an ideal case compared to two other graphs since both the methods [54, 66] approximate acceptable hemispheres compared to the baseline. One observes the blue and the green cuts in the rightmost graph which are very much close to the cut obtained from the baseline in red. In this case, the camera focal length was chosen in both of the methods [54, 66] to be 64. As the middle and the left graphs show, reduction of focal length to 32 and 16, respectively, reduces the accuracy of [66], namely the perspective geometry of the green cut gets far from the cut representing the baseline model

2.5 Summary and Conclusion

We reviewed perspective SFS techniques and recalled corresponding modeling assumptions. In addition, the FM algorithm as a computationally effective method was discussed in the context of a classic PSFS approach [66]. We compared the latter method with the approach of Tankus et al. [54] which is supposed to rely on the same modeling assumptions but which features a different implementation of the FM strategy by construction.

We could verify that the computational approach of Tankus and co-authors shows a stable and reliable behavior, especially for varying focal lengths and regimes that are perspective and close to the orthographic case. In contrast, we proposed a tractable algorithm for the method of Yuen and co-authors, which in general appears to give reasonable results, yet which has to be handled carefully if the input features a small focal length.

It appears at first glance a bit surprising that two methods that should be very close by the underlying model as well as by the underlying algorithm give very different results in some settings as one could observe by our study in this chapter. This shows once more, that in the field of SFS, one needs to be very careful in the design of models and algorithms. Moreover, a clean modeling by making use of a PDE, which is the basis of the model of Tankus et al., seems to bear not only theoretical but also practical advantages, as there are no hidden assumptions or parameters in this setting

Appendices

Appendix 1

The irradiance equation

$$I_0 = \frac{|z_0 A_3 + B_3|}{\sqrt{(z_0 A_1 + B_1)^2 + (z_0 A_2 + B_2)^2 + (z_0 A_3 + B_3)^2}}$$

is reformulated as

$$I_0^2 \left((z_0 A_1 + B_1)^2 + (z_0 A_2 + B_2)^2 + (z_0 A_3 + B_3)^2 \right) = (z_0 A_3 + B_3)^2$$

and further simplified as

$$
\begin{aligned}
I_0^2 (\cdots \\
\left(z_0^2 A_1^2 + B_1^2 + 2z_0 A_1 B_1 \right) + \\
\left(z_0^2 A_2^2 + B_2^2 + 2z_0 A_2 B_2 \right) + \\
\left(z_0^2 A_3^2 + B_3^2 + 2z_0 A_3 B_3 \right) \\
\cdots) = z_0^2 A_3^2 + B_3^2 + 2z_0 A_3 B_3.
\end{aligned}
\tag{2.45}
$$

By distributing the I_0^2, on rewrites (2.45) as

$$
\begin{aligned}
z_0^2 A_1^2 I_0^2 + B_1^2 I_0^2 + 2z_0 A_1 B_1 I_0^2 + \\
z_0^2 A_2^2 I_0^2 + B_2^2 I_0^2 + 2z_0 A_2 B_2 I_0^2 + \\
z_0^2 A_3^2 I_0^2 + B_3^2 I_0^2 + 2z_0 A_3 B_3 I_0^2 - \\
z_0^2 A_3^2 + B_3^2 + 2z_0 A_3 B_3 = 0.
\end{aligned}
\tag{2.46}
$$

Rearranging (2.46) while factoring out z_0^2 and z_0 results it to be written as

$$
\overbrace{\left(I_0^2 \left(A_1^2 + A_2^2 + A_3^2 \right) - A_3^2 \right)}^{C_1} z_0^2 +
$$

$$
\overbrace{\left(2I_0^2 \left(A_1 B_1 + A_2 B_2 + A_3 B_3 \right) - 2A_3 B_3 \right)}^{C_2} z_0 +
$$

$$\overbrace{\left(I_0^2 \left(B_1^2 + B_{\cdot\cdot}^2 + B_3^2\right) - B_3^2\right)}^{C_3} = 0, \tag{2.47}$$

and finally the image irradiance equation is reformulated in the form of the quadratic equation

$$C_1 z_0^2 + C_2 z_0 + C_3 = 0. \tag{2.48}$$

Appendix 2

In this case, the term $(P_a - P_0)^\top$ is redefined as

$$P_a - P_0 = \begin{pmatrix} v_b z_b - v_0 z_0 \\ -(u_b z_b - u_0 z_0) \\ 0 \end{pmatrix} \tag{2.49}$$

restricting the wave-front to only propagate from the direction of P_b, namely $(P_a - P_0) \cdot (P_b - P_0) = 0$, leading the derivation of the normal N to proceed as

$$\mathbf{N} = (P_a - P_0)^\top \times (P_b - P_0)^\top = \begin{pmatrix} v_b z_b - v_0 z_0 \\ -(u_b z_b - u_0 z_0) \\ 0 \end{pmatrix} \times \begin{pmatrix} \frac{u_b z_b - u_0 z_0}{f} \\ \frac{v_b z_b - v_0 z_0}{f} \\ z_b - z_0 \end{pmatrix}. \tag{2.50}$$

Now by letting

$$x_b := (u_b z_b - u_0 z_0) \quad \text{and} \quad y_b := v_b z_b - v_0 z_0$$

one writes (2.50) as

$$\mathbf{N} = (P_a - P_0)^\top \times (P_b - P_0)^\top$$
$$= \begin{pmatrix} y_b \\ -x_b \\ 0 \end{pmatrix} \times \begin{pmatrix} x_b/f \\ y_b/f \\ z_b - z_0 \end{pmatrix}$$
$$= \begin{pmatrix} -x_b (z_b - z_0) \\ -y_b (z_b - z_0) \\ \frac{x_b^2 + y_b^2}{f} \end{pmatrix} \tag{2.51}$$

as the normal vector to the surface point P_0 in case of the degenerated case $\eta_1 = +\infty$.

Appendix 3

Because (u_b, v_b) is a neighbor of (u_0, v_0), one can write x_b and y_b as

$$x_b = u_0 (z_b - z_0) + \Delta_1 z_b \qquad (2.52)$$

and

$$y_b = v_0 (z_b - z_0) + \Delta_2 z_b \qquad (2.53)$$

with $(\Delta_1, \Delta_2) \in \{(0, \pm 1), (\pm 1, 0)\}$, and substitute them into the numerator of the irradiance image

$$I_0 = \frac{x_b^2 + y_b^2}{\sqrt{f^2 x_b^2 (z_b - z_0)^2 + f^2 y_b^2 (z_b - z_0)^2 + \left(x_b^2 + y_b^2\right)^2}}$$

to get

$$I_0 = \frac{\overbrace{(u_0 (z_b - z_0) + \Delta_1 z_b)^2}^{x_b^2} + \overbrace{(v_0 (z_b - z_0) + \Delta_2 z_b)^2}^{y_b^2}}{\sqrt{f^2 x_b^2 (z_b - z_0)^2 + f^2 y_b^2 (z_b - z_0)^2 + \left(x_b^2 + y_b^2\right)^2}}$$

that is expanded as

$$I_0 = \frac{\left\{ \begin{array}{l} u_0^2 (z_b - z_0)^2 + \Delta_1^2 z_b^2 + 2u_0 (z_b - z_0) \Delta_1 z_b + \cdots \\ v_0^2 (z_b - z_0)^2 + \Delta_2^2 z_b^2 + 2v_0 (z_b - z_0) \Delta_2 z_b \end{array} \right\}}{\sqrt{f^2 x_b^2 (z_b - z_0)^2 + f^2 y_b^2 (z_b - z_0)^2 + \left(x_b^2 + y_b^2\right)^2}}. \qquad (2.54)$$

In addition, by factoring $f^2 (z_b - z_0)$ from the first two terms of the (2.54) denominator, we have

$$I_0 = \frac{\left\{ \begin{array}{l} u_0^2 (z_b - z_0)^2 + \Delta_1^2 z_b^2 + 2u_0 (z_b - z_0) \Delta_1 z_b + \cdots \\ v_0^2 (z_b - z_0)^2 + \Delta_2^2 z_b^2 + 2v_0 (z_b - z_0) \Delta_2 z_b \end{array} \right\}}{\sqrt{f^2 (z_b - z_0)^2 \left(x_b^2 + y_b^2\right) + \left(x_b^2 + y_b^2\right)^2}}$$

that is more simplified in its denominator as

$$I_0 = \frac{\left\{ \begin{array}{l} u_0^2 (z_b - z_0)^2 + \Delta_1^2 z_b^2 + 2u_0 (z_b - z_0) \Delta_1 z_b + \cdots \\ v_0^2 (z_b - z_0)^2 + \Delta_2^2 z_b^2 + 2v_0 (z_b - z_0) \Delta_2 z_b y_b \end{array} \right\}}{\sqrt{\left(x_b^2 + y_b^2\right) \left(f^2 (z_b - z_0)^2 + \left(x_b^2 + y_b^2\right)\right)}}. \qquad (2.55)$$

Taking both sides of (2.55) to the power of 2, we are lead to

$$
I_0^2 = \frac{\left\{ \begin{array}{l} u_0^2 \left(z_b - z_0\right)^2 + \Delta_1^2 z_b^2 + 2u_0 \left(z_b - z_0\right) \Delta_1 z_b + \cdots \\ v_0^2 \left(z_b - z_0\right)^2 + \Delta_2^2 z_b^2 + 2v_0 \left(z_b - z_0\right) \Delta_2 z_b \end{array} \right\}^2}{\left(x_b^2 + y_b^2\right) \left(f^2 \left(z_b - z_0\right)^2 + \left(x_b^2 + y_b^2\right)\right)} \equiv \left(x_b^2 + y_b^2\right)^2
$$

letting us to have the image irradiance equation as

$$
I_0^2 = \frac{\left\{ \begin{array}{l} u_0^2 \left(z_b - z_0\right)^2 + \Delta_1^2 z_b^2 + 2u_0 \left(z_b - z_0\right) \Delta_1 z_b + \cdots \\ v_0^2 \left(z_b - z_0\right)^2 + \Delta_2^2 z_b^2 + 2v_0 \left(z_b - z_0\right) \Delta_2 z_b \end{array} \right\}}{f^2 \left(z_b - z_0\right)^2 + \left(x_b^2 + y_b^2\right)}.
$$

Now, all terms are taken to the same side

$$
\left\{ \begin{array}{l} u_0^2 \left(z_b - z_0\right)^2 + \Delta_1^2 z_b^2 + 2u_0 \left(z_b - z_0\right) \Delta_1 z_b + \cdots \\ v_0^2 \left(z_b - z_0\right)^2 + \Delta_2^2 z_b^2 + 2v_0 \left(z_b - z_0\right) \Delta_2 z_b - \cdots \\ I_0^2 f^2 \left(z_b - z_0\right)^2 - I_0^2 \left(x_b^2 + y_b^2\right) \end{array} \right\} = 0
$$

and further simplified based on the common factor $\left(z_b - z_0\right)$ as

$$
\left\{ \begin{array}{l} \left(u_0^2 + v_0^2 - I_0^2 f^2\right) \left(z_b - z_0\right)^2 + \cdots \\ \left(2u_0 \Delta_1 z_b + 2v_0 \Delta_2 z_b\right) \left(z_b - z_0\right) + \cdots \\ \Delta_1^2 z_b^2 + \Delta_2^2 z_b^2 - I_0^2 \left(x_b^2 + y_b^2\right) \end{array} \right\} = 0. \tag{2.56}
$$

To this end, once again the terms x_b and y_b in (2.56) need to be replaced by (2.52) and (2.53) as

$$
\left\{ \begin{array}{l} \left(u_0^2 + v_0^2 - I_0^2 f^2\right) \left(z_b - z_0\right)^2 + \cdots \\ \left(2u_0 \Delta_1 z_b + 2v_0 \Delta_2 z_b\right) \left(z_b - z_0\right) + \cdots \\ \Delta_1^2 z_b^2 + \Delta_2^2 z_b^2 - I_0^2 \underbrace{\left\{ \begin{array}{l} \overbrace{u_0^2 \left(z_b - z_0\right)^2 + \Delta_1^2 z_b^2 + 2u_0 \left(z_b - z_0\right) \Delta_1 z_b + \cdots}^{x_b^2} \\ \underbrace{v_0^2 \left(z_b - z_0\right)^2 + \Delta_2^2 z_b^2 + 2v_0 \left(z_b - z_0\right) \Delta_2 z_b}_{y_b^2} \end{array} \right\}} \end{array} \right\} = 0
$$

and once again rearranged based on $\left(z_b - z_0\right)$ and $\left(z_b - z_0\right)^2$ as

$$
\left\{ \begin{array}{l} \left(\left(u_0^2 + v_0^2 - I_0^2 f^2\right) - I_0^2 \left(u_0^2 + v_0^2\right)\right) \left(z_b - z_0\right)^2 + \cdots \\ \left(2u_0 \Delta_1 z_b + 2v_0 \Delta_2 z_b - I_0^2 2u_0 \Delta_1 z_b - I_0^2 2v_0 \Delta_2 z_b\right) \left(z_b - z_0\right) + \cdots \\ \Delta_1^2 z_b^2 + \Delta_2^2 z_b^2 - I_0^2 \Delta_1^2 z_b^2 - I_0^2 \Delta_2^2 z_b^2 \end{array} \right\} = 0
$$

or

$$
\left\{
\begin{array}{l}
\left((u_0^2 + v_0^2 - I_0^2 f^2) - I_0^2 \left(u_0^2 + v_0^2\right)\right) (z_b - z_0)^2 + \cdots \\
\left(2u_0 \Delta_1 z_b \left(1 - I_0^2\right) + 2v_0 \Delta_2 z_b \left(1 - I_0^2\right)\right) (z_b - z_0) + \cdots \\
\Delta_1^2 z_b^2 \left(1 - I_0^2\right) + \Delta_2^2 z_b^2 \left(1 - I_0^2\right)
\end{array}
\right\} = 0
$$

that leads to

$$
\left\{
\begin{array}{l}
\overbrace{\left((u_0^2 + v_0^2 - I_0^2 f^2) - I_0^2 \left(u_0^2 + v_0^2\right)\right)}^{D_1} (z_b - z_0)^2 + \cdots \\
\overbrace{\left((2u_0 \Delta_1 z_b + 2v_0 \Delta_2 z_b) \left(1 - I_0^2\right)\right)}^{D_2} (z_b - z_0) + \cdots \\
\underbrace{\left(\Delta_1^2 z_b^2 + \Delta_2^2 z_b^2\right) \left(1 - I_0^2\right)}_{D_3}
\end{array}
\right\} = 0
$$

and finally written as a quadratic equation

$$
D_1 (z_b - z_0)^2 + D_2 (z_b - z_0) + D_3 = 0 \tag{2.57}
$$

with below coefficients:

$$
D_1 := \left(u_0^2 + v_0^2\right) - I_0^2 \left(f^2 + u_0^2 + v_0^2\right), \qquad D_2 := 2z_b \left(u_0 \Delta_1 + v_0 \Delta_2\right) \left(1 - I_0^2\right),
$$

$$
D_3 := \left(\Delta_1^2 z_b^2 + \Delta_2^2 z_b^2\right) \left(1 - I_0^2\right).
$$

Appendix 4

Steps in the direction of normal vector derivation by Tankus et al. [54]:

$$
\mathbf{n} = \left(\frac{d}{du} C_1(u)\right) \times \left(\frac{d}{dv} C_2(v)\right)
$$

$$
= \frac{1}{f}
\begin{pmatrix}
-z - u z_u \\
-v_0 z_u \\
f z_u
\end{pmatrix}
\times
\frac{1}{f}
\begin{pmatrix}
-u_0 z_v \\
-z - v z_v \\
f z_v
\end{pmatrix}
$$

$$
= \frac{1}{f^2}
\begin{pmatrix}
-z - u z_u \\
-v_0 z_u \\
f z_u
\end{pmatrix}
\times
\begin{pmatrix}
-u_0 z_v \\
-z - v z_v \\
f z_v
\end{pmatrix}
$$

$$
= \frac{1}{f^2}
\begin{pmatrix}
-v z_u \cdot f z_v + f z_u (z + v z_v) \\
-f z_u \cdot u z_v + (z + u z_u) f z_v \\
(z + u z_u)(z + v z_v) - v z_u \cdot u z_v
\end{pmatrix}
$$

$$= \frac{1}{f^2} \begin{pmatrix} -fvz_uz_v + fzz_u + fvz_uz_v \\ -fuz_uz_v + fzz_v + fuz_uz_v \\ z^2 + vzz_v + uzz_u + uvz_uz_v - uvz_uz_v \end{pmatrix}$$

$$= \frac{1}{f^2} \begin{pmatrix} fzz_u \\ fzz_v \\ z^2 + vzz_v + uzz_u \end{pmatrix}$$

$$= \frac{1}{f^2} \begin{pmatrix} fzz_u \\ fzz_v \\ z^2 + z\,(vz_v + uz_u) \end{pmatrix}$$

$$= \frac{z}{f^2} \begin{pmatrix} fz_u \\ fz_v \\ z + vz_v + uz_u \end{pmatrix}. \tag{2.58}$$

Based on (2.58), the unit normal vector is found as

$$\hat{\mathbf{n}} = \frac{\mathbf{n}}{\|\mathbf{n}\|}$$

$$= \frac{\frac{z}{f^2}\,(fz_u, fz_v, z + vz_v + uz_u)}{\sqrt{f^2 z_u^2 \frac{z^2}{f^4} + f^2 z_v^2 \frac{z^2}{f^4} + (z + vz_v + uz_u)^2 \frac{z^2}{f^4}}}$$

$$= \frac{\frac{z}{f^2}\,(fz_u, fz_v, z + vz_v + uz_u)}{\frac{z}{f^2}\sqrt{f^2 z_u^2 + f^2 z_v^2 + (z + vz_v + uz_u)^2}}$$

$$= \frac{(fz_u, fz_v, z + vz_v + uz_u)}{\sqrt{f^2\,(z_u^2 + z_v^2) + (z + vz_v + uz_u)^2}}. \tag{2.59}$$

Appendix 5

Steps to derive the image irradiance equation (2.37) proposed by Tankus et al. [54] and based on (2.34), (2.35) and (2.36).

$$I = \frac{(u - fp_s)\,z_u + (v - fq_s)\,z_v + z}{\|L\|\sqrt{f^2\,(z_u^2 + z_v^2) + (z + vz_v + uz_u)^2}}$$

$$= \frac{(u - fp_s)\,pz + (v - fq_s)\,qz + z}{\|L\|\sqrt{f^2\left(\underbrace{(pz)^2}_{z_u^2} + \underbrace{(qz)^2}_{z_v^2}\right) + (z + vqz + upz)^2}}$$

$$= \frac{(u - fp_s)\, pz + (v - fq_s)\, qz + z}{\|L\| \sqrt{f^2 \underbrace{\left(p^2 z^2 + q^2 z^2\right)}_{\beta} + (z + vqz + upz)^2}}$$

$$= \frac{(u - fp_s)\, pz + (v - fq_s)\, qz + z}{\|L\| \sqrt{f^2 z^2 \underbrace{\left(p^2 + q^2\right)}_{\beta} + (z + vqz + upz)^2}}$$

$$= \frac{(u - fp_s)\, pz + (v - fq_s)\, qz + z}{\|L\| \sqrt{f^2 z^2 \beta + (z + vqz + upz)^2}}$$

$$= \frac{(u - fp_s)\, pz + (v - fq_s)\, qz + z}{\|L\| \sqrt{f^2 z^2 \beta + \left(z^2 + v^2 q^2 z^2 + u^2 p^2 z^2 + 2vqz^2 + 2upz^2 + 2uvpqz^2\right)}}$$

$$= \frac{(u - fp_s)\, pz + (v - fq_s)\, qz + z}{\|L\| \sqrt{f^2 z^2 \beta + z^2 \left(1 + v^2 q^2 + u^2 p^2 + 2vq + 2up + 2uvpq\right)}}$$

$$= \frac{(u - fp_s)\, pz + (v - fq_s)\, qz + z}{\|L\| \sqrt{f^2 z^2 \beta + z^2 (up + vq + 1)^2}}$$

$$= \frac{z\left((u - fp_s)\, p + (v - fq_s)\, q + 1\right)}{z\|L\| \sqrt{f^2 \beta + (up + vq + 1)^2}}$$

$$= \frac{(u - fp_s)\, p + (v - fq_s)\, q + 1}{\|L\| \sqrt{f^2 \beta + (up + vq + 1)^2}}$$

$$= \frac{(u - fp_s)\, p + (v - fq_s)\, q + 1}{\|L\| \sqrt{f^2 \left(p^2 + q^2\right) + (up + vq + 1)^2}}$$

$$= \frac{(u - fp_s)\, p + (v - fq_s)\, q + 1}{\|L\| \sqrt{f^2 \left(p^2 + q^2\right) + (up + vq + 1)^2}}.$$

Appendix 6

Steps to further simplify the image irradiance equation (2.37) of Tankus et al. [54] to the form shown in (2.38) proceeds by letting both sides of (2.37) to the power of 2 as

$$I^2 = \frac{\left((u - fp_s)\, p + (v - fq_s)\, q + 1\right)^2}{\left(\|L\| \sqrt{f^2 \left(p^2 + q^2\right) + (up + vq + 1)^2}\right)^2}$$

and rearranging it as

$$I^2 \left(\|L\| \sqrt{f^2 \left(p^2 + q^2\right) + (up + vq + 1)^2} \right)^2 = ((u - fp_s)\, p + (v - fq_s)\, q + 1)^2 .$$

Taking all the terms to the left side

$$I^2 \left(\|L\| \sqrt{f^2 \left(p^2 + q^2\right) + (up + vq + 1)^2} \right)^2 - ((u - fp_s)\, p + (v - fq_s)\, q + 1)^2 = 0,$$

that simplifies to

$$\left(I^2 \|L\|^2 \left(f^2 \left(p^2 + q^2\right) + (up + vq + 1)^2 \right) \right) - ((u - fp_s)\, p + (v - fq_s)\, q + 1)^2 = 0,$$

by canceling the square root and finally appears as

$$\left(\underbrace{I^2 \|L\|^2 f^2 \left(p^2 + q^2\right)}_{\alpha_1} + \underbrace{I^2 \|L\|^2 (up + vq + 1)^2}_{\alpha_2} \right) - \underbrace{((u - fp_s)\, p + (v - fq_s)\, q + 1)^2}_{\alpha_3} = 0.$$

Appendix 7

The expanded forms of α_1, α_2 and α_3 are provided as:

– α_1:

$$\begin{aligned} \alpha_1 &= I^2 \|L\|^2 f^2 \left(p^2 + q^2\right) \\ &= I^2 \|L\|^2 f^2 p^2 + I^2 \|L\|^2 f^2 q^2 \end{aligned}$$

– α_2:

$$\begin{aligned} \alpha_2 &= I^2 \|L\|^2 (up + vq + 1)^2 \\ &= I^2 \|L\|^2 \left(u^2 p^2 + v^2 q^2 + 1 + 2pquv + 2up + 2vq\right) \\ &= I^2 \|L\|^2 u^2 p^2 + I^2 \|L\|^2 v^2 q^2 + I^2 \|L\|^2 + \cdots \\ &\quad 2I^2 \|L\|^2 pquv + 2I^2 \|L\|^2 up + 2I^2 \|L\|^2 vq \end{aligned}$$

– α_3:

$$\begin{aligned} \alpha_3 &= - \left((u - fp_s)\, p + (v - fq_s)\, q + 1 \right)^2 \\ &= -(u - fp_s)^2\, p^2 - (v - fq_s)^2\, q^2 - 1 - \cdots \\ &\quad - 2pq\, (u - fp_s)\, (v - fq_s) - 2p\, (u - fp_s) - 2q\, (v - fq_s) \end{aligned}$$

Appendix 8

To derive (2.39), let us start from (2.38) and proceed as

$$
\left\{
\begin{array}{l}
\underbrace{I^2\|L\|^2 f^2\left(p^2+q^2\right)+\cdots}_{\alpha_1} \\[2mm]
\underbrace{I^2\|L\|^2\left(up+vq+1\right)^2-\cdots}_{\alpha_2} \\[2mm]
\underbrace{\left(\left(u-fp_s\right)p+\left(v-fq_s\right)q+1\right)^2}_{\alpha_3}
\end{array}
\right\}=0.
$$

Now, those components have the terms of interest $p^2, q^2, 2pq, 2p$ and $2q$ in common which are marked as

$$
\left\{
\begin{array}{l}
I^2\|L\|^2 f^2 p^2+I^2\|L\|^2 f^2 q^2+\cdots \\
I^2\|L\|^2 u^2 p^2+I^2\|L\|^2 v^2 q^2+I^2\|L\|^2+2I^2\|L\|^2 pquv+2I^2\|L\|^2 up+2I^2\|L\|^2 vq \\
\cdots-\left(u-fp_s\right)^2 p^2-\left(v-fq_s\right)^2 q^2-1 \\
\cdots-2pq\left(u-fp_s\right)\left(v-fq_s\right)-2p\left(u-fp_s\right)-2q\left(v-fq_s\right)
\end{array}
\right\}=0,
$$

that leads to

$$
\left\{
\begin{array}{l}
p^2\overbrace{\left(I^2\|L\|^2\left(f^2+u^2\right)-\left(u-fp_s\right)^2\right)}^{:=A}+\cdots \\[3mm]
q^2\overbrace{\left(I^2\|L\|^2\left(f^2+v^2\right)-\left(v-fq_s\right)^2\right)}^{:=B}+\cdots \\[3mm]
2pq\overbrace{\left(I^2\|L\|^2 uv-\left(u-fp_s\right)\left(v-fq_s\right)\right)}^{:=C}+\cdots \\[3mm]
2p\overbrace{\left(I^2\|L\|^2 u-\left(u-fp_s\right)\right)}^{:=D}+\cdots \\[3mm]
2q\overbrace{\left(I^2\|L\|^2 v-\left(v-fq_s\right)\right)}^{:=E}+\cdots \\[3mm]
\overbrace{I^2\|L\|^2-1}^{:=F}
\end{array}
\right\}=0,
$$

and finally to

$$
p^2 A+q^2 B+2pqC+2pD+2qE+F=0.
$$

Algorithm 1 : Perspective fast marching, Yuen et al. [66].

Initialization

0. Create the depth map ζ, assign all of its node values to $+\infty$ and all of their labels *far*.

1. Locate the nodes in depth map corresponding to the boundary areas ∂I, label them as *known* and set their depth values to minimum depth value.

2. For each grid point $\zeta_{u,v}$ belonging to the depth map, define the neighbor sets

$$N_h = \left\{\zeta_{u-1,v}, \zeta_{u+1,v}\right\} \quad and \quad N_v = \left\{\zeta_{u,v-1}, \zeta_{u,v+1}\right\}.$$

3. Define the *narrow band* as the nodes belonging to $N_h(\partial I) \cup N_v(\partial I)$.

Update cycle

4. Take the node in narrow band with minimum depth value, call it κ, label it as *known* and remove it from the narrow band.

5. Look for all neighbors $N_h(\kappa) \cup N_v(\kappa)$. If they are not *known* label them as *trial* else if their labels are *far*, add them to the narrow band.

6. Find η_1 and η_2 such that

$$\eta_1 = \min\{N_h(\kappa)\} \quad and \quad \eta_2 = \min\{N_v(\kappa)\}.$$

7. Compute ζ_κ based on η_1 and η_2 as

if $(\eta_1 = +\infty \vee \eta_2 = +\infty)$ **then**
  if $(\eta_1 = +\infty)$ **then**
    Call Algorithm (**Degenerated horizontal case, Yuen et al. [66]**)
  else $\{(\eta_2 = +\infty)\}$
    Call Algorithm (**Degenerated vertical case, Yuen et al. [66]**)
  end if
else $\{(\eta_1 \neq +\infty \wedge \eta_2 \neq +\infty)\}$
  Call Algorithm (**Main Case, Yuen et al. [66]**)
end if

8. Stop if the *narrow band* is empty, else go to step **4**.

Algorithm 2 : Degenerated (horizontal/vertical) cases, Yuen et al. [66].
Note: In the vertical case, the subscripts a and b are interchanged while deriving the normal vector (2.60) in Appendix 2. Later, (2.61)–(2.63) are adjusted according to the subscript change and toward computation of the sought depth z_0.

0. Compute the normal $\mathbf{N}$ to the surface point P_κ as

$$\mathbf{N} = \left(-x_b \left(z_b - z_\kappa \right), \, -y_b \left(z_b - z_\kappa \right), \, \frac{x_b^2 + y_b^2}{f} \right)^{\top} \tag{2.60}$$

See Appendix 2 for details.

1. Rewrite the image irradiance equation (2.23) by adopting the normal $\mathbf{N}$ derived in (2.60) as

$$I_\kappa = \frac{x_b^2 + y_b^2}{\sqrt{f^2 x_b^2 \left(z_b - z_\kappa \right)^2 + f^2 y_b^2 \left(z_b - z_\kappa \right)^2 + \left(x_b^2 + y_b^2 \right)^2}}. \tag{2.61}$$

2. Rewrite the image irradiance equation (2.61) as quadratic equation

$$D_1 \left(z_b - z_\kappa \right)^2 + D_2 \left(z_b - z_\kappa \right) + D_3 = 0. \tag{2.62}$$

See Appendix 3 for details.

3. Obtain the depth ζ_κ as one of the following cases:

$$\begin{cases} \zeta_{u,v} := z_\kappa = z_b + \left(\left(-D_2 + \sqrt{D_2^2 - 4D_1 D_3} \right) / 2D_1 \right), & D \ge 0 \\ \zeta_{u,v} := z_\kappa = z_b + \left(\left(-D_2 - \sqrt{D_2^2 - 4D_1 D_3} \right) / 2D_1 \right), & D_1 < 0 \wedge D_2 < 0 \wedge D_4 \ge 0 \\ \zeta_{u,v} := z_\kappa = +\infty, & D_1 < 0 \wedge D_2 \ge 0 \wedge D_4 \ge 0 \\ \zeta_{u,v} := z_\kappa = +\infty, & D_4 < 0 \end{cases} \tag{2.63}$$

Algorithm 3 : Main case, Yuen et al. [66].

---

0. Rewrite the irradiance image Equation (2.23) in the form of a quadratic equation

$$C_1 z_\kappa^2 + C_2 z_\kappa + C_3 = 0 \quad \text{with} \quad C_4 := C_2^2 - 4C_1 C_3.$$

See Appendix 1 for details.

1. Let

$$Z_- := \min\left(-c_2 \pm \sqrt{c_4}/2c_1\right) \quad \text{and} \quad Z_+ := \max\left(-c_2 \pm \sqrt{c_4}/2c_1\right).$$

2. Obtain the depth ζ_κ as one of the following cases:

if $(C_4 < 0 \vee Z_+ = \max(\eta_1, \eta_2))$ **then**
   $\zeta_i = +\infty$ with $\zeta_i := \arg\max(\eta_1, \eta_2)$
   if $\eta_1 = +\infty$ **then**
      Call Algorithm (**Degenerated horizontal case, Yuen et al. [66]**)
   else $\{\eta_2 = +\infty\}$
      Call Algorithm (**Degenerated vertical case, Yuen et al. [66]**)
   end if
else

$$\begin{cases} \zeta_\kappa = Z_-, & Z_- \geq \max(\eta_1, \eta_2) \\ \zeta_\kappa = Z_+, & Z_+ \geq \max(\eta_1, \eta_2) > Z_- \\ \zeta_\kappa = Z_+, & Z_- < \max(\eta_1, \eta_2) \\ \zeta_\kappa = Z_-, & \text{otherwise} \end{cases}$$

end if

Algorithm 4 : Perspective fast marching, Tankus et al. [54].

Initialization

0. Create the depth map ζ, assign all of its node values to $+\infty$ and all of their labels *far*.

1. Locate the nodes in depth map corresponding to the boundary areas ∂I, label them as *known* and set their depth values to minimum depth value.

2. For each grid point $Z_{u,v}$ belonging to the depth map, define the neighbor sets

$$N_h = \left\{\zeta_{u-1,v}, \zeta_{u+1,v}\right\} \quad and \quad N_v = \left\{\zeta_{u,v-1}, \zeta_{u,v+1}\right\}.$$

3. Define the *narrow band* as the nodes belonging to $N_h(\partial I) \cup N_v(\partial I)$.

Update cycle

4. Take the node in narrow band with minimum depth value, call it κ, label it as *known* and remove it from the narrow band.

5. Look for all neighbors $N_h(\kappa) \cup N_v(\kappa)$. If they are not *known* label them as *trial* else if their labels are *far*, add them to the narrow band.

6. Find η_1 and η_2 such that

$$\eta_1 = \min\{N_h(\kappa)\} \quad and \quad \eta_2 = \min\{N_v(\kappa)\}.$$

7. Compute the ζ_κ based on η_1 and η_2 as

if $\left(\eta_2 - \eta_1 > \sqrt{F_\kappa/A_1}\right)$ **then**
  $\zeta_\kappa = \eta_1 + \sqrt{F_\kappa/A_1}$
else $\{\eta_1 - \eta_2 > \sqrt{F_\kappa/B_1}\}$
  $\zeta_\kappa = \eta_2 + \sqrt{F_\kappa/B_1}$
else $\{\eta_2 - \eta_1 \leq \sqrt{F_\kappa/A_1} \wedge \eta_1 - \eta_2 \leq \sqrt{F_\kappa/B_1}\}$
  $\zeta_\kappa = \dfrac{A_1\eta_1 + B_1\eta_2 \pm \sqrt{(A_1+B_1)\hat{F}_A - A_1 B_1(\eta_1-\eta_2)^2}}{A_1+B_1}$
end if

8. Stop if the *narrow band* is empty, else go to step **4.**

References

1. Abdelrahim AS, Abdelrahman MA, Abdelmunim H, Farag A, Miller M (2011) Novel image-based 3D reconstruction of the human jaw using shape from shading and feature descriptors. In: Proceedings of the British machine vision conference (BMVC), pp 1–11
2. Ahmed A, Farag A (2006) A new formulation for shape from shading for non-Lambertian surfaces. In: Proceedings of 2006 IEEE computer society conference on computer vision and pattern recognition, vol 2, June 2006. IEEE Computer Society Press, New York, NY, pp 17–22
3. Aloimonos JY (1990) Perspective approximations. Image Vis Comput 8(3):179–192
4. Bakshi S, Yang YH (1994) Shape from shading for non-Lambertian surfaces. In: Proceedings of IEEE international conference on image processing, vol 2, Austin, TX. IEEE Computer Society Press, pp 130–134

5. Bors AG, Hancock ER, Wilson RC (2003) Terrain analysis using radar shape-from-shading. IEEE Trans Pattern Anal Mach Intell 25(8):974–992
6. Breuß M, Cristiani E, Durou J-D, Falcone M, Vogel O (2012) Perspective shape from shading: ambiguity analysis and numerical approximations. SIAM J Imaging Sci 5(1):311–342
7. Cacace S, Cristiani E, Falcone M (2014) Can local single-pass methods solve any stationary Hamilton-Jacobi-Bellman equation? SIAM J Sci Comput 36(2):570–587
8. Camilli F, Tozza S (2017) A unified approach to the well-posedness of some non-Lambertian models in shape-from-shading theory. SIAM J Imaging Sci 10(1):26–46
9. Cho SI, Saito H, Ozawa S (1997) A divide-and-conquer strategy in shape from shading. In: Proceedings of the 1997 IEEE computer society conference on computer vision and pattern recognition (CVPR), June 17–19, San Juan, Puerto Rico, pp 413–419
10. Courteille F, Crouzil A, Durou J-D, Gurdjos P (2007) Shape from shading for the digitization of curved documents. Mach Vis Appl 18:301–316
11. Courteille F, Crouzil A, Durou J-D, Gurdjos P (2004) Shape from shading en conditions ralistes d'acquisition photographique. In: 14ème Congrs Francophone de Reconnaissance des Formes et Intelligence Artificielle - RFIA 2004, Toulouse, 28 January 2004–30 January 2004, vol 2, AFRIF-AFIA, pp 925–934
12. Courteille F, Crouzil A, Durou J-D, Gurdjos P (2004) Towards shape from shading under realistic photographic conditions. In: Proceedings of 17th international conference on pattern recognition, vol II, August 2004. Cambridge, UK, pp 277–280
13. Durou J-D, Falcone M, Sagona M (2008) Numerical methods for shape-from-shading: a new survey with benchmarks. Comput Vis Image Underst 109(1):22–43
14. Forsyth D (2011) Variable-source shading analysis. Int J Comput Vis 91(3):280–302
15. Galliani S, Ju YC, Breuß M, Bruhn A (2013) Generalised perspective shape from shading in spherical coordinates. In: Kuijper A, Pock T, Bredies K, Bischof H (eds) Proceedings of 4th international conference on scale space and variational methods in computer vision (SSVM 2013), Graz, Austria, June 2013. Lecture notes in computer science, vol 7893. Springer, Berlin, pp 222–233
16. Gårding J (1992) Shape from texture for smooth curved surfaces in perspective projection. J Math Imaging Vis 2:327–350
17. Gårding J (1993) Shape from texture and contour by weak isotropy. Artif Intell 64:243–297
18. Georgoulis S, Rematas K, Ritschel T, Gavves E, Fritz M, Van Gool L, Tuytelaars T (2018) Reflectance and natural illumination from single-material specular objects using deep learning. IEEE Trans Pattern Anal Mach Intell
19. Gibson JJ (1950) The perception of the visual world. Houghton Mifflin, Boston
20. Hartley R, Zisserman A (2000) Multiple view geometry in computer vision. Cambridge University Press, Cambridge
21. Hasegawa JK, Tozzi CL (1996) Shape from shading with perspective projection and camera calibration. Comput Graph 20(3):351–364
22. Helmsen JJ, Puckett EG, Colella P, Dorr M (1996) Two new methods for simulating photolithography development in 3d. In: Optical microlithography IX, Santa Clara, CA, USA, March 1996. SPIE, pp 253–261
23. Horn BKP, Brooks MJ (1989) Shape from shading. Artificial intelligence series. MIT Press, Cambridge
24. Horn BKP (1970) Shape from shading: a method for obtaining the shape of a smooth opaque object from one view. PhD thesis, Department of Electrical Engineering, MIT, Cambridge, Massachusetts, USA
25. Horn BKP (1975) Obtaining shape from shading information. In: Winston PH (ed) The psychology of computer vision. McGraw-Hill, New York, pp 115–155
26. Horn BKP (1986) Robot vision. MIT Press, Cambridge
27. Ju YC, Tozza S, Breuß M, Bruhn A, Kleefeld A (2013) Generalised perspective shape from shading with Oren-Nayar reflectance. In: Burghardt T, Damen D, Mayol-Cuevas W, Mirmehdi M (eds) Proceedings of 24th British machine vision conference (BMVC 2013), Bristol, UK, September 2013, Article 42. BMVA Press

28. Kimmel R, Bruckstein AM (1995) Tracking level sets by level sets: a method for solving the shape from shading problem. Comput Vis Image Underst 62(1):47–58
29. Lambert JH (1760) Photometria: sive de mensura et gradibus luminis, colorum et umbrae. Augsburg, sumptibus viduae E. Klett
30. Lee KM, Kuo CCJ (1994) Shape from shading with perspective projection. Comput Vis Graph Image Process: Image Underst 59(2):202–211
31. Lee K, Kuo CCJ (1997) Shape from shading with a generalized reflectance map model. Comput Vis Image Underst 67(2):143–160
32. Li Z, Xu Z, Ramamoorthi R, Sunkavalli K, Chandraker M (2018) Learning to reconstruct shape and spatially-varying reflectance from a single image. ACM Trans Graph 37(6):269:1–269:11
33. Maurer D, Ju YC, Breuß M, Bruhn A (2018) Combining shape from shading and stereo: a joint variational method for estimating depth, illumination and albedo. Int J Comput Vis 126(12):1342–1366
34. Okatani T, Deguchi K (1997) Shape reconstruction from an endoscope image by shape from shading technique for a point light source at the projection center. Comput Vis Image Underst 66(2):119–131
35. Oliensis J, Dupuis P (1991) Direct method for reconstructing shape from shading. In: Proceedings of SPIE conference 1570 on geometric methods in computer vision, San Diego, California, July 1991, pp 116–128
36. Oren M, Nayar SK (1995) Generalization of the Lambertian model and implications for machine vision. Int J Comput Vis 14(3):227–251
37. Penna MA (1989) Local and semi-local shape from shading for a single perspective image of a smooth object. Comput Vis Graph Image Process 46(3):346–366
38. Penna MA (1989) A shape from shading analysis for a single perspective image of a polyhedron. IEEE Trans Pattern Anal Mach Intell 11(6):545–554
39. Prados E, Camilli F, Faugeras O (2006) A unifying and rigorous shape from shading method adapted to realistic data and applications. J Math Imaging Vis 25(3):307–328
40. Prados E, Camilli F, Faugeras O (2006) A viscosity solution method for shape-from-shading without image boundary data. ESAIM: Math Model Numer Anal 40(2):393–412
41. Prados E, Faugeras O (2003) "Perspective shape from shading" and viscosity solutions. In: Proceedings of 9th IEEE international conference on computer vision, vol II, Nice, France, October 2003, pp 826–831
42. Prados E, Faugeras O (2004) Unifying approaches and removing unrealistic assumptions in shape from shading: mathematics can help. In: Proceedings of 8th European conference on computer vision, vol IV, Prague, Czech Republic, May 2004. LNCS, vol 3024, pp 141–154
43. Prados E, Faugeras O (2005) Shape from shading: a well-posed problem? In: Proceedings of IEEE conference on computer vision and pattern recognition, vol II, San Diego, California, USA, June 2005, pp 870–877
44. Ragheb H, Hancock E (2005) Surface radiance correction for shape-from-shading. Pattern Recognit 38(10):1574–1595
45. Rhodin H, Breuß M (2013) A mathematically justified algorithm for shape from texture. In: Kuijjper A, Pock T, Bredies K, Bischof H (eds) Proceedings of 4th international conference on scale space and variational methods in computer vision (SSVM 2013), Graz, Austria, June 2013. Lecture notes in computer science, vol 7893, pp 294–305. Springer, Berlin
46. Rindfleisch T (1966) Photometric method for lunar topography. Photogramm Eng 32(2):262–277
47. Samaras D, Metaxas D (2003) Incorporating illumination constraints in deformable models for shape from shading and light direction estimation. IEEE Trans Pattern Anal Mach Intell 25(2):247–264
48. Sethian JA (1996) Fast marching level set methods for three-dimensional photolithography development. In: Optical microlithography IX, Santa Clara, CA, USA, March 1996. SPIE, pp 262–272
49. Sethian JA (1999) Level set methods and fast marching methods, 2nd edn. Cambridge University Press, Cambridge. Paperback edition

50. Sethian JA (1996) A fast marching level set method for monotonically advancing fronts. Proc Natl Acad Sci 93(4):1591–1595
51. Smith WAP, Hancock ER (2008) Facial shape-from-shading and recognition using principal geodesic analysis and robust statistics. Int J Comput Vis 76(1):71–91
52. Tankus A, Sochen N, Yeshurun Y (2005) Shape-from-shading under perspective projection. Int J Comput Vis 63(1):21–43
53. Tankus A, Sochen N, Yeshurun Y (2003) A new perspective [on] shape-from-shading. In: Proceedings of 9th IEEE international conference on computer vision, vol II, Nice, France, October 2003, pp 862–869
54. Tankus A, Sochen N, Yeshurun Y (2004) Perspective shape-from-shading by fast marching. In: Proceedings of IEEE conference on computer vision and pattern recognition, vol I, Washington, D.C., USA, June 2004, pp 43–49
55. Tozza S, Falcone M (2016) Analysis and approximation of some shape-from-shading models for non-Lambertian surfaces. J Math Imaging Vis 55(2):153–178
56. Tsitsiklis JN (1995) Efficient algorithms for globally optimal trajectories. IEEE Trans Autom Control 40(9):1528–1538
57. Van Diggelen J (1951) A photometric investigation of the slopes and heights of the ranges of hills in the Maria of the moon. Bull Astron Inst Neth XI(423):283–289
58. Vogel O, Breuß M, Weickert J (2008) Perspective shape from shading with non-Lambertian reflectance. In: Rigoll G (ed) Pattern recognition. Lecture notes in computer science, vol 5096. Springer, Berlin, pp 517–526
59. Wöhler C, Grumpe A (2013) Integrated DEM reconstruction and calibration of hyperspectral imagery: a remote sensing perspective. In: Breuß M, Bruckstein A, Maragos P (eds) Innovations for shape analysis: models and algorithms. Mathematics and visualization. Springer, Berlin, pp 467–492
60. Woodham RJ (1980) Photometric method for determining surface orientation from multiple images. Opt Eng 19(1):139–144
61. Wu C, Narasimhan S, Jaramaz B (2010) A multi-image shape-from-shading framework for near-lighting perspective endoscopes. Int J Comput Vis 86(2):211–228
62. Wu J, Smith WAP, Hancock ER (2010) Facial gender classification using shape from shading. Image Vis Comput 28(6):1039–1048
63. Yamany SM, Farag AA, Tasman D, Farman AGP (1999) A robust 3-D reconstruction system for human jaw modeling. In: Taylor C, Colchester A (eds) Proceedings of 2nd international conference on medical image computing and computer-assisted intervention (MICCAI'99), Cambridge, England, September 1999. LNCS, vol 1679, pp 778–787
64. Yamany SM, Farag AA, Tasman D, Farman AGP (2000) A 3-D reconstruction system for the human jaw using a sequence of optical images. IEEE Trans Med Imaging 11(5):538–547
65. Yuen SY, Tsui YY, Leung YW, Chen RMM (2002) Fast marching method for shape from shading under perspective projection. In: Proceedings of 2nd international conference visualization, imaging, and image processing (VIIP 2002), Marbella, Spain, September 2002, pp 584–589
66. Yuen SY, Tsui YY, Chow CK (2007) A fast marching formulation of perspective shape from shading under frontal illumination. Pattern Recognit Lett 28:806–824
67. Zhang R, Tsai P-S, Cryer JE, Shah M (1999) Shape from shading: a survey. IEEE Trans Pattern Anal Mach Intell 21(8):690–706

Chapter 3
RGBD-Fusion: Depth Refinement for Diffuse and Specular Objects

Roy Or-El, Elad Richardson, Matan Sela, Rom Hershkovitz, Aaron Wetzler, Guy Rosman, Alfred M. Bruckstein, and Ron Kimmel

Abstract The popularity of low-cost RGB-D scanners is increasing on a daily basis and has set off a major boost in 3D computer vision research. Nevertheless, commodity scanners often cannot capture subtle details in the environment. In other words, the precision of existing depth scanners is often not accurate enough to recover fine details of scanned objects. In this chapter, we review recent axiomatic methods to enhance the depth map by fusing the intensity and depth information to create detailed range profiles. We present a novel shape-from-shading framework that enhances the quality of recovery of diffuse and specular objects' depth profiles. The first shading-based depth refinement method we review is designed to work well with Lambertian objects, however, it breaks down in the presence of specularities. To that end, we propose a second method, which utilizes the properties of the built-in monochromatic IR projector and the acquired IR images of common RGB-D scanners and propose a lighting model that accounts for the specular regions in the input image. In the methods suggested above, the detailed geometry is calculated without the need to explicitly find and integrate surface normals, this allows the numerical implementations to work in real-time. Finally, we also show how we can leverage deep learning to refine depth details. We present a neural network that is trained with the above models and can be naturally integrated as part of a larger network architecture. Both quantitative tests and visual evaluations prove that the suggested methods produce state-of-the-art depth reconstruction results.

R. Or-El (✉)
University of Washington, Seattle, WA, USA
e-mail: royorel@cs.washington.edu

E. Richardson · M. Sela · R. Hershkovitz · A. Wetzler · A. M. Bruckstein · R. Kimmel
Technion, Israel Institute of Technology, Haifa, Israel

G. Rosman
Computer Science and Artificial Intelligence Lab, MIT, Cambridge, MA, USA

© Springer Nature Switzerland AG 2020
J.-D. Durou et al. (eds.), *Advances in Photometric 3D-Reconstruction*,
Advances in Computer Vision and Pattern Recognition,
https://doi.org/10.1007/978-3-030-51866-0_3

3.1   Introduction

The availability of affordable depth scanners has sparked a revolution in many applications of computer vision, such as robotics, human motion capture, and scene modeling and analysis. The increased availability of such scanners naturally raises the
question of whether it is possible to exploit the combined intensity and InfraRed (IR)
images to overcome their lack of accuracy. It is clear that to obtain fine details such
as fine facial features, one must compensate for the measurement errors inherent in
modern depth scanners.

Our goal is to fuse the captured data from the RGB-D scanner in order to enhance
the accuracy of the acquired depth maps. For this purpose, we should precisely align
and combine both depth and scene color or intensity cues. Assuming that the scanner
is stationary and its calibration parameters are known, aligning the intensity and
depth data is a relatively straightforward task. Recently, scanners that allow access
to both infrared scene illumination and depth maps, have become available enabling
the possibility of even richer RGB-D-I fusion.

Reconstructing a shape from color or intensity images, a task known as shape-
from-shading [1–3], is a well-addressed area in computer vision. These shape estimation solutions often suffer from ambiguities since there can be several possible
surfaces that can explain a given image. Recently, attempts have been made to eliminate some of these ambiguities by using more elaborated lighting models, with richer,
natural illumination environments [4, 5]. Moreover, it was observed that data from
depth sensors combined with shape-from-shading methods can be used to eliminate
ambiguities and improve the reconstructed depth maps [6–8].

Here, we discuss two recent real-time methods to directly enhance surface recovery that achieve state-of-the-art accuracy. The first approach applies a lighting model
that uses normals estimated from the depth profile and eliminates the need for calibration of the scene lighting. The lighting model accounts for distant light sources,
multiple albedos, and some local lighting effects. The result is a depth recovery
method that works well with Lambertian surfaces. However, most surfaces are not
purely Lambertian and exhibit specular reflections which may affect the depth refinement result. To that end, we adopt a second approach that uses the IR image supplied
by modern depth scanners to deal with specular reflections. The narrow-band nature
of the IR projector and IR camera provides a controlled lighting environment. We
then exploit this friendly environment to introduce a new depth refinement process
based on a lighting model that accounts for specular reflections as well as multiple
albedos.

Assuming that the lighting model explains the smooth nature of the intensity
image, and that high frequency data in the image is related to the surface geometry,
we reconstruct a high quality surface without explicitly finding and integrating its
normals. Instead, we use the relation between the surface gradient, its normals, and a
smoothed version of the input depth map to define a surface dependent cost functional.
In order to achieve fast convergence, we iteratively linearize the variational problem.

Additionally, we also discuss FineNet, a CNN architecture for accurate surface recovery. As the methods above, this network takes an intensity image and an initial depth profile as inputs and fuses shading cues from the input image to generate an enhanced depth map containing subtle details. This network is trained in an unsupervised manner using a loss function inspired by the above methods.

This chapter combines and streamlines the methods described in [9, 10], with the FineNet framework presented in [11]. The main topics addressed in this chapter are

1. Presenting a novel and robust depth enhancement method that operates under natural illumination and handles multiple albedos and specular objects.
2. Showing that depth accuracy can be improved in a real-time system by efficiently fusing the RGB-D/IR-D inputs.
3. Demonstrating that improved depth maps can be acquired directly using shape-from-shading technique that avoids the need to first find the surface normals and then integrate them.
4. Introducing an unsupervised CNN framework for depth enhancement, based on the presented axiomatic SfS model.

The chapter outline is as follows: we overview previous and related efforts in Sect. 3.2. The proposed algorithms are presented in Sects. 3.3, 3.4 and 3.5. Results are shown in Sect. 3.6, with discussions in Sect. 3.7.

3.2 Related Work

We first briefly review some of the research done in depth enhancement and shape-from-shading. We refer to just a few representative papers that capture the major development and the state-of-the-art in these fields.

3.2.1 Depth Enhancement

Depth enhancement algorithms mostly rely on one of the following strategies: using multiple depth maps, employing pre-learned depth priors and combining depth and intensity maps.

Multiple depth maps. Chen and Medioni, laid the foundation to this paradigm in [12] by registering overlapping depth maps to create an accurate and complete 3D models of objects. Digne et al. [13] decomposed laser scans to low and high frequency components using the intrinsic heat equation. They fuse together the low frequency components of the scans and keep the high frequency data untouched to produce a higher resolution model of an object. Merrel et al. [14] generated depth images from intensity videos which were later fused to create a high resolution 3D model of objects. Schuon et al. [15] aligned multiple slightly translated depth maps to enhance depth resolution. They later extended this in [16] to shape reconstruction from the global alignment of several super-resolved depth maps. Tong et al. [17] used a nonrigid registration technique to combine depth videos from three Kinects to produce a high resolution scan of a human body. Probably the most popular effort in this area is the KinectFusion algorithm [18], in which a real-time depth stream is fused on the GPU into a truncated signed distance function to accurately describe a 3D model of a scanned volume. In [19] Maier et al. extend the KinectFusion model and jointly optimize the SDF along with shading cues from the RGB images. Zou et al. [20] refine the object's depth by exploiting the lighting variations from casual movement to calculate photometric stereo under natural illumination. Sang et al. [21] propose to attach an LED light to an RGB-D scanner and reconstruct the surface from multiple viewpoints where the motion is unknown.

Pre-learned depth priors. Oisin et al. [22] use a dictionary of synthetic depth patches to build a high resolution depth map. Hornáček et al. [23] extended to 3D both the self similarity method introduced in [24] and the PatchMatch algorithm by Barnes et al. [25] and showed how they can be coupled to increase spatial and depth resolution. Li et al. [26] extract features from a training set of high resolution color image and low resolution depth map patches, then, they learn a mapping function between the color and low resolution depth patches to the high resolution depth patches. Finally, depth resolution is enhanced by a sparse coding algorithm. In the context of 3D scanning, Rosman et al. [27] demonstrated the use of a sparse dictionary for range images for 3D structured-light reconstruction.

Depth and intensity maps. The basic assumption behind these methods is that depth discontinuities are strongly related to intensity discontinuities. In [28, 29], a joint bilateral upsampling of intensity images was used to enhance the depth resolution. Park et al. [30] combined a nonlocal means regularization term with an edge weighting neighborhood smoothness term and a data fidelity term to define an energy function whose minimization recovers a high quality depth map. In a more recent paper, Lee and Lee [31], used an optical flow like algorithm to simultaneously increase the resolution of both intensity and depth images. This was achieved using a single depth image and multiple intensity images from a video camera. Lu et al. [32] assemble similar RGBD patches into a matrix and use its low-rank estimation to enhance the depth map.

3.2.2 Shape-from-Shading

Classical Shape-from-shading. The shape-from-shading problem under the assumption of a Lambertian surface and uniform illumination was first introduced by Horn in [1]. The surface is recovered using the characteristic strip expansion method. In 1986, Horn and Brooks [33], explored variational approaches for solving the shape-from-shading problem. Later, Bruckstein [2], developed a direct method of recovering the surface by level sets evolution assuming that the light source is directly above the surface. This method was later generalized by Kimmel and Bruckstein [34], to handle general cases of uniform lighting from any direction. In [3], Kimmel and

Sethian show that a fast solution to the shape from shading problem can be obtained from a modification to the fast marching algorithm. The works of Mecca et al. [35, 36], has recently resulted in a PDE formulation for direct surface reconstruction using photometric stereo. Two surveys in [37, 38], comprehensively cover the shape-from-shading problem as studied over the last few decades.

Recently, attempts were made to solve the shape-from-shading problem under uncalibrated natural illumination. Forsyth [39] modeled the shading by using a spatially varying light source to reconstruct the surface. Huang and Smith [4] use first-order spherical harmonics to approximate the surface reflectance map. The shape is then recovered by using an edge preserving smoothing constraint and by minimizing the local brightness error. Johnson and Adelson [5] modeled the shading as a quadratic function of the surface normals. They showed, counter-intuitively, that natural illumination reduces the surface normals ambiguity, and thus, makes the shape-from-shading problem simpler to solve. Queau et al. [40] propose a new PDE-based model for shape-from-shading, which handles various illumination and camera models. They also introduce a new variational scheme for solving the PDE-based model.

The main practical drawback about classical shape-from-shading is that although a diffusive single albedo setup can be easily designed in a laboratory, it can rarely be found in more realistic environments. As such, modern SfS approaches attempt to reconstruct the surface without any assumptions about the scene lighting and/or the object albedos. In order to account for the unknown scene conditions, these algorithms either use learning techniques to construct priors for the shape and scene parameters, or acquire a rough depth map from a 3D scanner to initialize the surface.

Learning-based methods. Barron and Malik [41] constructed priors from statistical data of multiple images to recover the shape, albedo and illumination of a given input image. Kar et al. [42] learn 3D deformable models from 2D annotations in order to recover detailed shapes. Richter and Roth [43] extract color, textons and silhouette features from a test image to estimate a reflectance map from which patches of objects from a database are rendered and used in a learning framework for regression of surface normals. Although these methods produce excellent results, they depend on the quality and size of their training data, whereas the proposed axiomatic approach does not require a training stage and is therefore applicable in more general settings.

Depth map-based methods. Bohme et al. [44] find a MAP estimate of an enhanced range map by imposing a shading constraint on a probabilistic image formation model. Zhang et al. [6] fuse depth maps and color images captured under different illumination conditions and use photometric stereo to improve the shape quality. Yu et al. [7] use mean shift clustering and second-order spherical harmonics to estimate the depth map scene albedos and lighting from a color image. These estimations are then combined together to improve the given depth map accuracy. Han et al. [45] propose a quadratic global lighting model along with a spatially varying local lighting model to enhance the quality of the depth profile. Kadambi et al. [46] fuse normals obtained from polarization cues with rough depth maps to obtain accurate reconstructions. Even though this method can handle specular surfaces, it requires at least three photos to reconstruct the normals and it does not run in real-time. Several IR-based methods were introduced in [8, 47–49]. The authors of [8, 48] suggest a multi-shot photometric stereo approach to reconstruct the object normals. Choe et al. [47] refine 3D meshes from Kinect Fusion [18] using IR images captured during the fusion pipeline. Although this method can handle uncalibrated lighting, it is neither one-shot nor real-time since a mesh must first be acquired before the refinement process begins. Ti et al. [49] propose a simultaneous time-of-flight and photometric stereo algorithm that utilizes several light sources to produce accurate surface and surface normals. Although this method can be implemented in real time, it requires four shots per frame for reconstruction as opposed to our single-shot approach. More inline with our approach, Wu et al. [50] use second order spherical harmonics to estimate the global scene lighting, which is then followed by an efficient scheme to reconstruct the object. In [51], Haefner et al. use shape-from-shading to alleviate the ill-posedness of a single image super resolution and vice versa to enhance the resolution of a given depth map. A similar solution using an RGBD sequence and photometric stereo is proposed in [52]. While some shape-from-shading methods directly optimize for depth, many methods recover normals first, and then integrate them. The survey [53] summarizes recent advances in this field.

Multi-view Reconstruction. A recent approach in computer vision is to combine shape-from-shading with multi-view stereo to enhance the overall reconstruction quality. Langguth et al. [54] use the magnitude of image gradients to balance between a geometric and shading energy terms to reconstruct a surface given multiple views. Wu et al. [55] design a generative adversarial network that transforms multiple views of specular objects to diffuse ones, thus, enabling more accurate reconstruction by downstream shading-based algorithms. Guo et al. [56] propose a shading-based scheme that uses 3D data to estimate motion between two RGB-D frames. The recovered motion along with a volumetric albedo fusing scheme then used for refining the 3D geometry. Alternating between these two schemes yields a detailed 3D-shape, albedo and motion. Queau et al. [57] couple PDE-based shape-from-shading solutions for single image across multiple images and multiple color channels by means of a variational formulation to reconstruct a 3D shape given multiple views. In [58], Liu et al. propose a unified framework accounting for shape, shading and specularities for nonrigid 3D reconstruction given sequential RGB frames from a single camera.

3.3 Diffuse Shape Refinement

We next introduce a framework for depth refinement of diffuse objects. The input is a depth map and a corresponding intensity image. We assume that the input depth and intensity images were taken from a fixed calibrated system. The intrinsic matrices and the extrinsic parameters of the depth and color sensors are assumed to be known.

We first wish to obtain a rough version of the input surface. However, due to measurement inaccuracies, the given depth profile is fairly noisy. For a smooth estimate,
we apply a bilateral filter on the input depth map.

Next, we estimate initial surface normals corresponding to the smoothed surface.
A lighting model can now be evaluated. We start by recovering the shading from the
initial normals and the intensity. The subsequent step accounts for different albedos
and shadows. Finally, the last step estimates spatially varying illumination that better
explains local lighting effects which only affect portions of the image. Once the
lighting model is determined, we move on to enhance the surface. Like modern
shape-from-shading methods, we take advantage of the fact that an initial depth map
is given. With a depth map input, a high quality surface can be directly reconstructed
without first refining the estimated normals. This is done by a variational process
designed to minimize a depth-based cost functional. Finally, we show how to speed
up the reconstruction process.

3.3.1 Lighting Estimation

The shading function relates a surface geometry to its intensity image. The image is
taken under natural illumination where there is no single point light source. Thus, the
correct scene lighting cannot be recovered with a directional lighting model. A more
complex lighting model is needed. Grosse et al. [59] introduced an extended intrinsic image decomposition model that has been widely used for recovering intrinsic
images. We show how we can efficiently incorporate this model for our problem in
order to get state of the art surface reconstruction. Define,

$$L(i, j, \vec{n}) = \rho(i, j)S(\vec{n}) + \beta(i, j), \tag{3.1}$$

where $L(i, j, \vec{n})$ is the image irradiance at each pixel, $S(\vec{n})$ is the shading, $\rho(i, j)$
accounts for multiple scene albedos and shadowed areas since it adjusts the shading
intensity. $\beta(i, j)$ is added as an independent, spatially varying light source, that
accounts for local lighting variations such as interreflections and specularities. We
note that the (i, j) indexing is sometimes omitted for convenience throughout the
paper.

Clearly, since we only have a single input image, without any prior knowledge,
recovering S, ρ and β for each pixel is an ill-posed problem. However, the given
depth map helps us recover all three components for each pixel.

3.3.1.1 Shading Computation

First, we assume a Lambertian scene and a single directional light source and recover
the shading S, associated with light sources that have a uniform effect on the image.
Once the shading is computed, we move on to find ρ and β, to better explain the

intensity image given the object geometry. During the shading recovery process, we set ρ to 1 and β to 0.

Basri and Jacobs [60] and Ramamoorthi and Hanrahan [61] found that the irradiance of diffuse objects in natural illumination scenes can be well described by low order spherical harmonics components. Thus, a smooth function is sufficient to recover the shading image. For the sake of simple and efficient modeling, we opt to use zero and first-order spherical harmonics, which are a linear polynomial of the surface normals and are independent on the pixel's location. Therefore, they are given by

$$S(\vec{n}) = \vec{m}^T \tilde{n}, \tag{3.2}$$

where $\vec{n}$ is the surface normal, $S(\vec{n})$ is the shading function, $\vec{m}$ is a vector of the four first-order spherical harmonics coefficients, and $\tilde{n} = (\vec{n}, 1)^T$.

Every valid pixel in the aligned intensity image I can be used to recover the shading. Hence, we have an overdetermined least squares parameter estimation problem

$$\underset{\vec{m}}{\mathrm{argmin}} \, \|\vec{m}^T \tilde{n} - I\|_2^2. \tag{3.3}$$

The rough normals we obtained from the initial depth map eliminate the need for assumptions and constraints on the shape or using several images. This produces a straightforward parameter fitting problem unlike the classical shape-from-shading and photometric stereo approaches. Despite having only the normals of the smoothed surface we can still obtain an accurate shading model since the least square process is not sensitive to high frequency changes and subtle shape details. In addition, the estimated surface normals eliminate the need for pre-calibrating the system lighting and we can handle dynamic lighting environments.

Background normals obviously affect the shading model outcome since they are related to different materials with different albedos, hence, their irradiance is different. Nonetheless, unlike similar methods, our method is robust to such outliers as our lighting model and surface refinement scheme was designed to handle precisely that case.

3.3.1.2 Multiple Albedo Recovery

The shading alone gives us only a rough assessment of the lighting, as it explains mostly distant and ambient light sources and only holds for diffuse surfaces with uniform albedo. Specularities, shadows and nearby light sources remain unaccounted for. In addition, multiple scene albedos, unbalanced lighting or shadowed areas affect the shading model by biasing its parameters. An additional cause for the errors is the rough geometry used to recover the shading model in (3.2). In order to handle these problems, ρ and β should be computed.

Finding ρ and β is essential to enhance the surface geometry, without them, lighting variations will be incorrectly compensated for by adjusting the shape structure. Since we now have the shading S, we can move on to recover ρ.

Now, we freeze S to the shading image we just found and optimize ρ to distinguish between the scene albedos and account for shadows (β is still set to 0). We set a fidelity term to minimize the ℓ_2 error between the proposed model and the input image. However, without regularization, $\rho(i, j)$ is prone to overfitting since one can simply set $\rho = I/S$ and get an exact explanation for the image pixels. To avoid overfitting, a prior term that prevents ρ from changing rapidly is used. Thereby, the model explains only lighting changes and not geometry changes. We follow the retinex theory [62], and like other intrinsic images recovery algorithms, we assume that the albedo map is piecewise smooth and that there is a low number of albedos in the image. Unlike many intrinsic image recovery frameworks like [63, 64], who use a Gaussian mixture model for albedo recovery we use a weighted Laplacian to distinguish between materials and albedos on the scene while maintaining the smooth changing nature of light. This penalty term is defined as

$$\left\| \sum_{k \in \mathcal{N}} \omega_k^c \omega_k^d (\rho - \rho_k) \right\|_2^2, \tag{3.4}$$

where $\mathcal{N}$ is the neighborhood of the pixel. ω_k^c is an intensity weighting term suggested in Eq. (6), in [45]

$$\omega_k^c = \begin{cases} 0, & \|I_k - I\|_2^2 > \tau \\ \exp\left(-\frac{\|I_k - I(i,j)\|_2^2}{2\sigma_c^2}\right), & \text{otherwise,} \end{cases} \tag{3.5}$$

and ω_k^d is the following depth weighting term

$$\omega_k^d = \exp\left(-\frac{\|z_k - z(i,j)\|_2^2}{2\sigma_d^2}\right). \tag{3.6}$$

Here, σ_d is a parameter responsible for the allowed depth discontinuity and $z(i, j)$ represents the depth value of the respected pixel. This regularization term basically performs a three dimensional segmentation of the scene, dividing it into piecewise smooth parts. Therefore, material and albedo changes are accounted for but subtle changes in the surface are smoothed. To summarize, we have the following regularized linear least squares problem with respect to ρ

$$\min_{\rho} \|\rho S(\vec{n}) - I\|_2^2 + \lambda_\rho \| \sum_{k \in \mathcal{N}} \omega_k^c \omega_k^d (\rho - \rho_k)\|_2^2. \tag{3.7}$$

3.3.1.3 Lighting Variations Recovery

Finally, after $\rho(i, j)$ is found we move on to finding $\beta(i, j)$. A similar functional to the one used for $\rho(i, j)$ can also be used to recover $\beta(i, j)$, since specularities still maintain smooth variations. Despite that, we need to keep in mind the observation of [60, 61], first order spherical harmonics account for 87.5% of the scene lighting. Hence, we also limit the energy of $\beta(i, j)$ in order to be consistent with the shading model. Therefore, $\beta(i, j)$ is found by solving

$$\min_{\beta} \|\beta - (I - \rho S(\vec{n}))\|_2^2 + \lambda_\beta^1 \| \sum_{k \in \mathcal{N}} \omega_k^c \omega_k^d (\beta - \beta_k)\|_2^2 + \lambda_\beta^2 \|\beta\|_2^2. \tag{3.8}$$

3.3.2 Refining the Surface

At this point, our complete lighting model is set to explain the scene's lighting. Now, in order to complete the recovery process, fine geometry details need to be restored. A typical SFS method would now adjust the surface normals, trying to minimize

$$\|L(i, j, \vec{n}) - I\|_2^2 \tag{3.9}$$

along with some regularization terms or constraints. The resulting cost function will usually be minimized in the $(p - q)$ gradient space.

However, according to [39], in order to minimize (3.9) schemes that use the $(p - q)$ gradient space can yield surfaces that tilt away from the viewing direction. Moreover, an error in the lighting model which can be caused by normal outliers such as background normals would aggravate this artifact. Therefore, to avoid these phenomena, we take further advantage of the given depth map. We write the problem as a functional of z, and force the surface to change only in the viewing direction, limiting the surface distortion and increasing the robustness of the method to lighting model errors.

We use the geometric relation between surface normals and the surface gradient given by

$$\vec{n} = \frac{(z_x, z_y, -1)}{\sqrt{1 + \|\nabla z\|^2}}, \tag{3.10}$$

where

$$z_x = \frac{dz}{dx}, \quad z_y = \frac{dz}{dy}, \tag{3.11}$$

to directly enhance the depth map. The surface gradient, represented as a function of z, connects between the intensity image and the lighting model. Therefore, by fixing the lighting model parameters and allowing the surface gradient to vary, subtle details

---

Algorithm 1: Accelerated Surface Enhancement

Input: $z_0, \vec{m}, \rho, \beta$—initial surface, lighting parameters
1 **while** $f(z^{k-1}) - f(z^k) > 0$ **do**
2 | Update $\tilde{n}^k = (\vec{n}^k, 1)^T$
3 | Update $L(\nabla z^k) = \rho(\vec{m}^T \tilde{n}^k) + \beta$
4 | Update z^k to be the minimizer of $f(z^k)$
5 **end**

---

in the surface geometry can be recovered by minimizing the difference between the measured intensity image and our shading model,

$$\|L(\nabla z) - I\|_2^2. \tag{3.12}$$

Formulating the shape-from-shading term as function of z simplifies the numerical scheme and reduces ambiguities. Since we already have the rough surface geometry, only simple fidelity and smoothness terms are needed to regularize the shading. Therefore, our objective function for surface refinement is

$$f(z) = \|L(\nabla z) - I\|_2^2 + \lambda_z^1 \|z - z_0\|_2^2 + \lambda_z^2 \|\Delta z\|_2^2, \tag{3.13}$$

where z_0 is the initial depth map and Δ represents the Laplacian of the surface. Using a depth based numerical scheme instead of $(p - q)$ gradient space scheme, makes our algorithm less sensitive to noise and more robust to lighting model errors caused by normal outliers. This has a great implication in handling real-world scenarios where the desired shape cannot be easily distinguished from its background.

The functional introduced is nonlinear due to the shading term since the dependency between the surface normals and its gradient requires geometric normalization. A solution to (3.13) can be found using the Levenberg–Marquardt algorithm or various Trust-Region methods, however, their convergence is slow and not suitable for real-time applications.

In order to accelerate the performance of the algorithm, we reformulate the problem in a similar way to IRLS optimization scheme. We do so by freezing nonlinear terms inside the shading model. This allows us to solve a linear system at each iteration, and update the nonlinear terms at the end of each iteration. First, we recall Eq. (3.10) and eliminate the denumerator using the auxiliary variables

$$\vec{n}^k = w^k(z_x^k, z_y^k, -1)^T, \tag{3.14}$$

$$w^k = (1 + \|\nabla z^{k-1}\|^2)^{-\frac{1}{2}}$$

The new lighting linearized model reads

$$L(i, j, \nabla z) = \rho(i, j) \cdot (\vec{m}^T \tilde{n}^k) + \beta(i, j). \tag{3.15}$$

This results in updated shading term. Now, at each iteration we need to solve the following functional for z^k,

$$f(z^k) = \|\rho(\vec{m}^T \tilde{n}^k) - (I - \beta)\|_2^2 \\ + \lambda_z^1 \|z^k - z_0\|_2^2 + \lambda_z^2 \|\Delta z^k\|_2^2. \tag{3.16}$$

This process is repeated as long as the objective function $f(z)$ decreases. A detailed explanation of the update rule is displayed in Algorithm 1.

While the scheme is very efficient, it may fail to converge on some degenerate cases, for instance, for frontal lighting, when $l = (0, 0, -1)$. However, in natural illumination scenarios, such cases are very unlikely to occur. To guarantee convergence, one can use a Taylor approximation of the nonlinear terms at each iteration. Such a scheme is proposed in Sect. 3.4.3.

3.4   From Diffuse to Specular Surfaces

Shape-from-Shading (SfS) tries to relate an object's geometry to its image irradiance. Like many other inverse problems, SfS is also an ill-posed one because the per-pixel image intensity is determined by several elements: the surface geometry, its albedo, scene lighting, the camera parameters and the viewing direction.

When using depth maps from RGB-D scanners one could recover the camera parameters and viewing direction, yet, in order to obtain the correct surface, we first need to account for the scene lighting and the surface albedos. Failing to do so would cause the algorithm to change the surface geometry and introduce undesired deformations. Using cues from an RGB image under uncalibrated illumination like [9, 45, 50] requires an estimation of global lighting parameters. Although such estimations work well for diffuse objects, they usually fail when dealing with specular ones and result in a distorted geometry. The reason is that specularities are sparse outliers that are not accounted for by classical lighting models. Furthermore, trying to use estimated lighting directions to model specularities is prone to fail when there are multiple light sources in the scene.

In our scenario, the main lighting in the IR image comes from the scanner's projector, which can be treated as a point light source. Observe that in this setting, we do not need to estimate a global lighting direction, instead, we use a near light field model to describe the per-pixel lighting direction. Subsequently, we can also account for specularities and nonuniform albedo map. Note that unlike the distant light model assumed in the previous section, point source near field light model requires a perspective camera model.

Recall that, an initial depth estimation is given by the scanner. We avoid the process of computing a refined normal field and then fusing depth with normal estimates, which is common to SfS methods, and solve directly for the depth. This eliminates the need to enforce integrability and reduces the problem size by half. We deal with the nonlinear part by calculating a first-order approximation of the cost functional and thereby achieve real-time performance.

3.4.1   Specular Shape Refinement

A novel IR-based real-time framework for depth enhancement is proposed. The suggested algorithm requires a depth map and an IR image as inputs. We assume that the IR camera and the depth camera have the same intrinsic parameters, as is usually the case with common depth scanners. In addition, we also assume that the whole system is calibrated and that the translation vector between the scanner's IR projector and IR camera is known.

Unfortunately, the raw depth map is usually quantized and the surface geometry is highly distorted. Therefore, we first smooth the raw depth map and estimate the surface normals. We then move on to recover the scene lighting using a near-field lighting model which explicitly accounts for object albedos and specularities.

After we find the scene lighting along with albedo and specular maps, we can directly update the surface geometry by designing a cost functional that relates the depth and IR intensity values at each pixel. We also show how the reconstruction process can be accelerated in order to obtain real-time performance. Figure 3.1 shows a flowchart of the proposed algorithm.

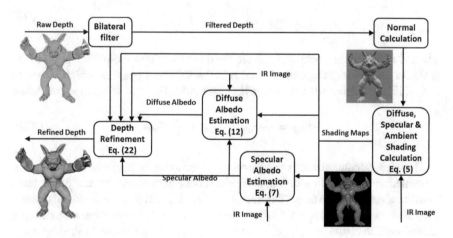

Fig. 3.1   Algorithm's flowchart

3.4.2 Near Field Lighting Model

Using an IR image as an input provides several advantages to the reconstruction process. Unlike other methods which require alignment between RGB and depth images, in our case, the depth map and IR image are already aligned as they were captured by the same camera. Moreover, the narrow-band nature of the IR camera means that the main light source in the image is the scanner's own IR projector whose location relative to the camera is known. Therefore, we can model the IR projector as a point light source and use a near field lighting model to describe the given IR image intensity at each pixel,

$$I = \frac{a\rho_d}{d_p^2} S_{diff} + \rho_d S_{amb} + \frac{a\rho_s}{d_p^2} S_{spec}. \tag{3.17}$$

Here, a is the projector intensity which is assumed to be constant throughout the image. d_p is the distance of the surface point from the projector. ρ_d and ρ_s are the diffuse and specular albedos. S_{amb} is the ambient lighting in the scene, which is also assumed to be constant over the image. S_{diff} is the diffuse shading function of the image which is given by the Lambertian reflectance model

$$S_{diff} = \vec{n} \cdot \vec{l}_p. \tag{3.18}$$

The specular shading function S_{spec} is set according to the Phong reflectance model

$$S_{spec} = \left(\left(2(\vec{l}_p \cdot \vec{n})\vec{n} - \vec{l}_p \right) \cdot \vec{l}_c \right)^\alpha, \tag{3.19}$$

where $\vec{n}$ is the surface normal, $\vec{l}_p, \vec{l}_c$ are the directions from the surface point to the projector and camera respectively and α is the shininess constant which we set to $\alpha = 2$. Figure 3.2 describes the scene lighting model. For ease of notation, we define

$$\tilde{S}_{diff} = \frac{a}{d_p^2} S_{diff}, \quad \tilde{S}_{spec} = \frac{a}{d_p^2} S_{spec}. \tag{3.20}$$

The intrinsic camera matrix and the relative location of the projector with respect to the camera are known. In addition, the initial surface normals can be easily calculated from the given rough surface. Therefore, $\vec{l}_c, \vec{l}_p, d_p, S_{diff}$ and S_{spec} can be found directly whereas a, S_{amb}, ρ_d and ρ_s need to be recovered. Although we are using a rough depth normal field to compute $\vec{l}_c, \vec{l}_p, d_p, S_{diff}$ and S_{spec} we still get accurate shading maps since the lighting is not sensitive to minor changes in the depth or normal field as shown in [60, 61]. Decomposing the IR image into its Lambertian and Specular lighting components along with their respective albedo maps has no unique solution. To achieve accurate results while maintaining real-time performance we choose a greedy approach which first assumes Lambertian lighting and gradually accounts for the lighting model from Eq. 3.17. Every pixel in the IR image which has an assigned

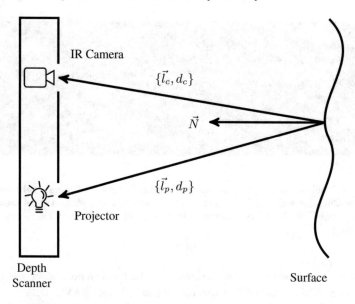

Fig. 3.2 Scene lighting model

normal can be used to recover a and S_{amb}. Generally, most of the light reflected back to the camera is related to the diffuse component of the object whereas highly specular areas usually have a more sparse nature. Thus, the specular areas can be treated as outliers in a parameter fitting scheme as they have minimal effect on the outcome. This allows us to assume that the object is fully Lambertian (i.e $\rho_d = 1$, $\rho_s = 0$), which in turn, gives us the following overdetermined linear system for n valid pixels ($n \gg 2$),

$$
\begin{pmatrix} \frac{S^1_{diff}}{(d^1_p)^2} & 1 \\ \vdots & \vdots \\ \frac{S^n_{diff}}{(d^n_p)^2} & 1 \end{pmatrix} \begin{pmatrix} a \\ S_{amb} \end{pmatrix} = \begin{pmatrix} I_1 \\ \vdots \\ I_n \end{pmatrix}.
\tag{3.21}
$$

3.4.2.1 Specular Albedo Map

The specular shading map is important since it reveals the object areas which are likely to produce specular reflections in the IR image. Without it, bright diffuse objects can be mistaken for specularities. Yet, since $\tilde{S}_{spec}$ was calculated as if the object is purely specular, using it by itself will fail to correctly represent the specular irradiance, as it would falsely brighten non-specular areas. In order to obtain an accurate representation of the specularities it is essential to find the specular albedo map to attenuate the non-specular areas of $\tilde{S}_{spec}$.

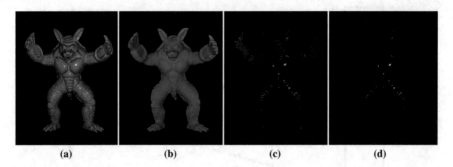

(a)   (b)   (c)   (d)

Fig. 3.3 **a** Simulated IR image of the Armadillo mesh. **b** Recovered image of the diffuse and ambient shading $\tilde{S}_{diff} + S_{amb}$. **c** Residual image for specular albedo estimation I_{res}^s. **d** Ground Truth specularity map of (**a**). Note that specularities in (**d**) are basically the sparse representation of the residual image (**c**)

We now show how we can take advantage of the sparse nature of the specularities to recover ρ_s and get the correct specular scene lighting. We will define a residual image I_{res}^s as being a difference between the original image I and our current diffuse approximation together with the ambient lighting. Formally, we write this as

$$I_{res}^s = I - (\tilde{S}_{diff} + S_{amb}). \tag{3.22}$$

As can be seen in Fig. 3.3c, the sparse bright areas of I_{res}^s are attributable to the true specularities in I. Specular areas have finite local support, therefore we choose to model the residual image I_{res}^s as $\rho_s \tilde{S}_{spec}$ such that ρ_s will be a sparse specular albedo map. This will yield an image that contains just the bright areas of I_{res}^s. In addition, in order to preserve the smooth nature of specularities we add a smoothness term that minimizes the L1 Total-Variation of ρ_s. To summarize, the energy minimization problem to estimate ρ_s can be written as

$$\min_{\rho_s} \lambda_1^s \|\rho_s \tilde{S}_{spec} - I_{res}^s\|_2^2 + \lambda_2^s \|\rho_s\|_1 + \lambda_3^s \|\nabla \rho_s\|_1, \tag{3.23}$$

where $\lambda_1^s, \lambda_2^s, \lambda_3^s$ are weighting terms for the fidelity, sparsity and smoothness terms, respectively. To minimize the cost functional, we use a variation of the Augmented Lagrangian method suggested in [65] where we substitute the frequency domain solution with a Gauss–Seidel scheme on the GPU. We refer the reader to the above paper for additional details on the optimization procedure.

3.4.2.2 Recovering the Diffuse Albedo

As was the case with specular shading, the diffuse shading map alone does not sufficiently explain the diffuse lighting. This is due to the fact that the diffuse shading is calculated as if there was only a single object with uniform albedo. In reality, however, most objects are composed of multiple different materials with different reflectance properties that need to be accounted for.

Using the estimated specular lighting from Sect. 3.4.2.1 we can now compute a residual image between the original image I and the specular scene lighting which we write as

$$I_{res}^d = I - \rho_s \tilde{S}_{spec}. \tag{3.24}$$

I_{res}^d should now contain only the diffuse and ambient irradiance of the original image I. This can be used in a data fidelity term for a cost functional designed to find the diffuse albedo map ρ_d.

We also wish to preserve the piecewise smoothness of the diffuse albedo map. Otherwise, geometry distortions will be mistaken for albedos and we will not be able to recover the correct surface. The IR image and the rough depth map provide us several cues that will help us to enforce piecewise smoothness. Sharp changes in the intensity of the IR image imply a change in the material reflectance. Moreover, depth discontinuities can also signal possible changes in the albedo.

We now wish to fuse the cues from the initial depth profile and the IR image together with the piecewise smooth albedo requirement. Past papers [9, 45] have used bilateral smoothing. Here, instead, we base our scheme on the geometric Beltrami framework such as in [66–68] which has the advantage of promoting alignment of the embedding space channels. Let,

$$\mathcal{M}(x, y) = \{x, y, \beta_I I_{res}^d(x, y), \beta_z z(x, y), \beta_\rho \rho_d(x, y)\} \tag{3.25}$$

be a two-dimensional manifold embedded in a $5D$ space with the metric

$$G = \begin{pmatrix} \langle \mathcal{M}_x, \mathcal{M}_x \rangle & \langle \mathcal{M}_x, \mathcal{M}_y \rangle \\ \langle \mathcal{M}_x, \mathcal{M}_y \rangle & \langle \mathcal{M}_y, \mathcal{M}_y \rangle \end{pmatrix}. \tag{3.26}$$

The gradient of ρ_d with respect to the $5D$ manifold is

$$\nabla_G \rho_d = G^{-1} \cdot \nabla \rho_d, \tag{3.27}$$

By choosing large enough values of β_I, β_z and β_ρ and minimizing the L1 Total-Variation of ρ_d with respect to the manifold metric, we basically perform selective smoothing according to the "feature" space (I_{res}^d, z, ρ_d). For instance, if $\beta_I \gg \beta_z, \beta_\rho, 1$, the manifold gradient would get small values when sharp edges are present in I_{res}^d since G^{-1} would decrease the weight of the gradient at such locations.

To conclude, the minimization problem we should solve in order to find the diffuse albedo map is

$$\min_{\rho_d} \lambda_1^d \left\| \rho_d \left(\tilde{S}_{diff} + S_{amb} \right) - I_{res}^d \right\|_2^2 + \lambda_2^d \| \nabla_G \rho_d \|_1. \tag{3.28}$$

Here, λ_1^d, λ_2^d are weighting terms for the fidelity and piecewise smooth penalties. We can minimize this functional using the Augmented Lagrangian method proposed in [69]. The metric is calculated separately for each pixel, therefore, it can be implemented very efficiently on a GPU with limited effect on the algorithm's runtime.

3.4.3 Surface Reconstruction

Once we account for the scene lighting, any differences between the IR image and the image rendered with our lighting model are attributed to geometry errors of the depth profile. Usually, shading-based reconstruction algorithms opt to use the dual stage process of finding the correct surface normals and then integrating them in order to obtain the refined depth. Although this approach is widely used, it has some significant shortcomings. Calculating the normal field is an ill-posed problem with $2n$ unknowns if n is the number of pixels. The abundance of variables can result in distorted surfaces that are tilted away from the camera. In addition, since the normal field is an implicit surface representation, further regularization such as the integrability constraint is needed to ensure that the resulting normals would represent a valid surface. This additional energy minimization functional can impact the performance of the algorithm.

Instead, we use the strategy suggested in [9, 50], and take advantage of the rough depth profile acquired by the scanner. Using the explicit depth values forces the surface to move only in the direction of the camera rays, avoids unwanted distortions, eliminates the need to use an integrability constraint and saves computation time and memory by reducing the number of variables.

In order to directly refine the surface, we relate the depth values to the image intensity through the surface normals. Assuming that the perspective camera intrinsic parameters are known, the $3D$ position $P(i, j)$ of each pixel is given by

$$P\left(z(i, j)\right) = \left(\frac{j - c_x}{f_x} z(i, j), \frac{i - c_y}{f_y} z(i, j), z(i, j)\right)^T, \qquad (3.29)$$

where f_x, f_y are the focal lengths of the camera and (c_x, c_y) is the camera's principal point. The surface normal $\vec{n}$ at each $3D$ point is then calculated by

$$\vec{n}\left(z(i, j)\right) = \frac{P_x \times P_y}{\|P_x \times P_y\|}. \qquad (3.30)$$

We can use Eqs. (3.17), (3.18) and (3.30) to write down a depth based shading term written directly in terms of z,

$$E_{sh}(z) = \left\| \frac{a\rho_d}{d_p^2} \left(\vec{n}(z) \cdot \vec{l}_p\right) + \rho_d S_{amb} + \rho_s \tilde{S}_{spec} - I \right\|_2^2. \qquad (3.31)$$

This allows us to refine z by penalizing shading mismatch with the original image I. We also use a fidelity term that penalizes the distance from the initial 3D points

$$E_f(z) = \|w(z - z_0)\|_2^2,$$

$$w = \sqrt{1 + \left(\frac{j - c_x}{f_x}\right)^2 + \left(\frac{i - c_y}{f_y}\right)^2}, \tag{3.32}$$

and a smoothness term that minimizes the second order TV-L1 of the surface

$$E_{sm}(z) = \|Hz\|_1, \quad H = \begin{pmatrix} D_{xx} \\ D_{yy} \end{pmatrix}. \tag{3.33}$$

Here, D_{xx}, D_{yy} are the second derivatives of the surface.

Combining Eqs. (3.31), (3.32) and (3.33) into a cost functional results in a non-linear optimization problem

$$\min_z \lambda_1^z E_{sh}(z) + \lambda_2^z E_f(z) + \lambda_3^z E_{sm}(z), \tag{3.34}$$

where $\lambda_1^z, \lambda_2^z, \lambda_3^z$ are the weights for the shading, fidelity and smoothness terms, respectively. Although there are several possible methods to solve this problem, a fast scheme is required for real-time performance. To accurately and efficiently refine the surface we base our approach on the iterative scheme suggested in [70]. Rewriting Eq. (3.31) as a function of the discrete depth map z, and using forward derivatives we have

$$I_{i,j} - \rho_d S_{amb} - \rho_s \tilde{S}_{spec} = \frac{a\rho_d}{d_p^2}(\vec{n}(z) \cdot \vec{l}_p)$$

$$= f(z_{i,j}, z_{i+1,j}, z_{i,j+1}). \tag{3.35}$$

At each iteration k we can approximate f using the first-order Taylor expansion about $(z_{i,j}^{k-1}, z_{i+1,j}^{k-1}, z_{i,j+1}^{k-1})$, such that

$$I_{i,j} - \rho_d S_{amb} - \rho_s \tilde{S}_{spec} = f(z_{i,j}^k, z_{i+1,j}^k, z_{i,j+1}^k)$$

$$\approx f(z_{i,j}^{k-1}, z_{i+1,j}^{k-1}, z_{i,j+1}^{k-1}) + \frac{\partial f}{\partial z_{i,j}^{k-1}}(z_{i,j}^k - z_{i,j}^{k-1})$$

$$+ \frac{\partial f}{\partial z_{i+1,j}^{k-1}}(z_{i+1,j}^k - z_{i+1,j}^{k-1}) + \frac{\partial f}{\partial z_{i,j+1}^{k-1}}(z_{i,j+1}^k - z_{i,j+1}^{k-1}). \tag{3.36}$$

Rearranging terms to isolate terms including z from the current iteration, we can define

$$I_{res}^{z^k} = I_{i,j} - \rho_d S_{amb} - \rho_s \tilde{S}_{spec}$$

$$- f(z_{i,j}^{k-1}, z_{i+1,j}^{k-1}, z_{i,j+1}^{k-1}) + \frac{\partial f}{\partial z_{i,j}^{k-1}} z_{i,j}^{k-1}, \qquad (3.37)$$

$$+ \frac{\partial f}{\partial z_{i+1,j}^{k-1}} z_{i+1,j}^{k-1} + \frac{\partial f}{\partial z_{i,j+1}^{k-1}} z_{i,j+1}^{k-1}$$

and therefore minimize

$$\min_{z^k} \lambda_1^z \|Az^k - I_{res}^{z^k}\|_2^2 + \lambda_2^z \|w(z^k - z_0)\|_2^2 + \lambda_3^z \|Hz^k\|_1 \qquad (3.38)$$

at each iteration with the Augmented Lagrangian method of [65]. Here, A is a matrix that represents the linear operations performed on the vector z^k. Finally, we note that this pipeline was implemented on an Intel i7 3.4 GHz processor with 16GB of RAM and an NVIDIA GeForce GTX690 GPU. The runtime for a 640×480 image is approximately 25 milliseconds.

3.5 Shape Refinement Learning

Traditionally, shape-from-shading is used in an online optimization framework where the shading term is minimized with respect to a given intensity image and, in our case, initial depth map. In some cases, the initial depth map can be directly inferred from the input image. For instance, Richardson et al. [71], used CoarseNet, a CNN architecture that iteratively regress facial 3D Morphable Model (3DMM) parameters from a single input image [72]. Unfortunately, this model cannot recover subtle facial features such as wrinkles that are not captured by a 3DMM model. One optional way to overcome this is to use the algorithm proposed in Sect. 3.3 as a post-processing step with the regressed face as the depth input as done in [71]. While the above pipeline produces compelling results this process can be streamlined by learning the entire shape-from-shading process. To that end, we propose FineNet, a CNN architecture for enhancing fine depth details. We note that FineNet is presented in the context of 3D face reconstruction but it can be used as a general platform for other depth enhancement applications.

3.5.1 FineNet Architecture

FineNet is based on the hypercolumn architecture suggested in [73]. The architecture's core idea is generating a feature map for each pixel that contains both structural and semantic data. This is done by concatenating the output responses from multiple layer outputs of the network. Due to pooling layers, the output maps size of inner

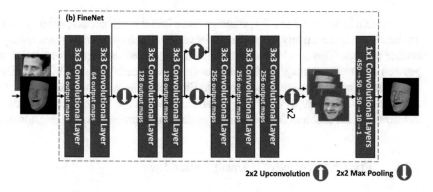

Fig. 3.4 FineNet architecture

layers is smaller than the size of the input image, therefore, we upscale them back to the original size, in order to create a dense volume of features of a size of the input image. This volume is then processed by several 1×1 convolution layers.

We use the VGG-Face [74] as a base for our hypercolumn network since it was pretrained on a domain of faces. Instead of directly upsampling each feature map to the original size using bilinear interpolation like in [73], we use cascaded 2-strided 2×2 transposed convolution layers. This results in improved features, as the interpolation is now also part of the learning process. Since refining the facial features is a relatively local problem, we truncate the VGG-Face network before the third pooling layer and form a $200 \times 200 \times 450$ hypercolumn feature volume. This volume is then processed by a set of 1×1 convolutional layers used as a linear regressor to create the final depth prediction. Note, that this fully convolutional framework allows us to use any size of input images. Figure 3.4 describes the FineNet architecture.

3.5.2 FineNet Unsupervised Loss

To train FineNet a loss function is required. One possible solution would be to simply use an MSE criterion between the network output and a high quality ground truth depth map. This would allow the network to implicitly learn how to reconstruct detailed faces from a single image. Unfortunately, a large dataset of detailed facial geometries with their corresponding 2D images is currently unavailable. Furthermore, a synthetic dataset for this task cannot be generated using morphable models as there is no known model that captures the diversity of fine facial details. Instead, we propose an unsupervised learning process where the loss criterion is determined by an axiomatic model. To achieve that, we need yet again to find a measure that relates the output depth map to the 2D image. To that end, we resort to Shape-from-Shading (SfS).

As shown in Sects. 3.3 and 3.4, when given an initial rough surface, subtle geometry details can be accurately recovered under various lighting conditions and multiple
surface albedos. This is achieved by optimizing an objective function which ties the
geometry to the input image. In this case, an initial surface is produced by CoarseNet
and its depth map representation is fed into FineNet along with the input image. We
then formulate an unsupervised loss criterion based on the SfS objective function,
transforming the problem from an online optimization problem to a regression one.

3.5.2.1   From SfS Objective to Unsupervised Loss

We formulate our unsupervised loss criterion in the spirit of Sect. 3.3. The main term
loss function is an image formation term, which models the connection between the
network's output depth map and the input image. That term drives the network to
recover the fine geometry details. It is defined as

$$E_{sh} = \left\| \rho \left\langle \vec{l}, \vec{Y}(\hat{z}) \right\rangle - I \right\|_2^2, \tag{3.39}$$

where $\hat{z}$ is the reconstructed depth map, I is the input intensity image, ρ is the albedo
image, and $\vec{l}$ are the first-order spherical harmonics coefficients. $Y(\hat{z})$ represents the
matching spherical harmonics basis,

$$Y(\hat{z}) = \left(1, n_x(\hat{z}), n_y(\hat{z}), n_z(\hat{z}) \right), \tag{3.40}$$

$(n_x(\hat{z}), n_y(\hat{z}), n_z(\hat{z}))$ denote the normal vector, expressed as a function of the depth.
Notice that the scene lighting $\vec{l}$ and albedo map ρ are unknown. Usually, recovering
both lighting and albedo causes ambiguity in SfS problems. In our case, we utilize
the fact that our problem is constrained to human faces. This limits the space of
possible albedo solutions to a subspace of low dimensional 3DMM texture.

$$\rho \approx T = \mu_T + A_T \alpha_T. \tag{3.41}$$

where μ_T is the average face texture, A_T is a principal component basis and α_T is the
corresponding coefficients vector. We used 10 coefficients in our implementation.

As shown in [75], by assuming the average facial albedo $\hat{\rho} = \mu_T$, the global
lighting can be correctly recovered using the coarse depth map, z_0, as follows:

$$\vec{l}^* = \underset{\vec{l}}{\operatorname{argmin}} \left\| \hat{\rho} \left\langle \vec{l}, \vec{Y}(z_0) \right\rangle - I \right\|_2^2. \tag{3.42}$$

Since this is an over determined linear problem, we can easily find the lighting
coefficients using least squares. Using the solution for the lighting, the albedo can
also be easily recovered as

Fig. 3.5 Light and albedo recovery. Images are presented next to the recovered albedo, rendered with the recovered lighting

$$\alpha_T^* = \operatorname*{argmin}_{\alpha_T} \left\| (\mu_T + A_T \alpha_T) \left\langle \vec{l}^*, \vec{Y}(z_0) \right\rangle - I \right\|_2^2. \tag{3.43}$$

Like Eq. (3.42), this over determined linear problem that can also be solved with least squares. Once the albedo and lighting coefficients are found, we can calculate E_{sh} and its gradient with respect to $\hat{z}$. A few recovery samples are presented in Fig. 3.5.

We add fidelity and smoothness terms to regularize the solution of FineNet. Namely,

$$\begin{aligned} E_f &= \|\hat{z} - z_0\|_2^2, \\ E_{sm} &= \|\Delta \hat{z}\|_1, \end{aligned} \tag{3.44}$$

where Δ is the discrete Laplacian operator. These regularizing terms guarantee the solution's smoothness and that it would not deviate from the prediction of CoarseNet. The overall loss function is

$$L(\hat{z}, z_0, I) = \lambda_{sh} E_{sh}(\hat{z}, I) + \lambda_f E_f(\hat{z}, z_0) + \lambda_{sm} E_{sm}(\hat{z}). \tag{3.45}$$

where the λ's determine the balance between the terms and in our experiments were set to $\lambda_{sh} = 1$, $\lambda_f = 5e^{-3}$, $\lambda_{sm} = 1$. The gradient of L with respect to $\hat{z}$ is then calculated and used for back-propagation.

3.5.2.2   Unsupervised Loss—a Discussion

Applying the unsupervised criterion has some desired benefits. It eliminates the need for an annotated dataset and also ensures that the network is not limited by the performance of any algorithm or the quality of the dataset. This is because the loss function is only dependent on the input, in contrast to supervised learning SfS schemes such as [76–78], where the data is generated by either photometric stereo, raw Kinect scans, or synthetic 3DMM models, respectively. In addition, since the albedo and lighting coefficients are calculated only as part of the loss function, the network is capable of producing accurate results directly from the intensity and depth inputs. Unlike traditional SfS algorithms, there is no need to explicitly calculate the albedo and lighting information. Although we can generate the lighting and albedo parameters as outputs of CoarseNet, we chose not to include them in the pipeline

Fig. 3.6 Criterion flow.
Gradients from both loss
criteria are propagated back
to CoarseNet

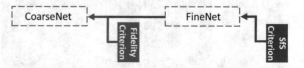

for two reasons. First, the lighting and albedo are only needed for the training stage
and have no use during testing. Second, both (3.42) and (3.43) are overdetermined
systems which can be solved efficiently with least squares, thus, using a network for
this task would be redundant.

3.5.3 End-to-End Network Training

Our CoarseNet was pre-trained on synthetic data to reconstruct the geometry captured by the 3DMM. Then, we connect the SfS FineNet to the CoarseNet through a
differential rendering component, and train the network as a whole using the criterion presented in Sect. 3.5.2 as the loss. Figure 3.6 describes the end-to-end training
scheme. Images from the VGG face dataset [74], were used for the end-to-end training. We refer the reader to [11, 71], for details about CoarseNet, the rendering layer,
and the face reconstruction problem.

3.6 Results

3.6.1 Diffuse Shape Refinement

In order to test the proposed algorithm, we performed a series of experiments to
validate its efficiency and accuracy. We show that our results are quantitatively and
visually state-of-the-art, using both synthetic and real data. In addition, we display
the ability of our algorithm to avoid texture copy artifacts, handle multiple albedo
objects, demonstrate the robustness of our algorithm to background normals outliers,
and present a runtime profile of the proposed method.

First, we start by performing a quantitative comparison between our method and
our implementation of the methods proposed by [45, 50], which will be referred to
as HLK and WZNSIT, respectively. In this experiment we use synthetic data in order
to have a reference model. We took objects from the Stanford 3D repository [79] and
the Smithsonian 3D archive and simulated a complex lighting environment using
Blender. In addition, we also used complete 3D models and scenes from the public
Blendswap repository. Each model is used to test a different scenario, "Thai Statue"[1]
tests a Lambertian object in a three-point lighting environment with minimal shad-

---

[1]http://graphics.stanford.edu/data/3Dscanrep.

Table 3.1 Quantitative comparison of depth accuracy on simulated models

	Median				90th%			
	Initial	Han et al.	Wu et al.	Proposed	Initial	Han et al.	Wu et al.	Proposed
Thai Statue	1.014	0.506	0.341	**0.291**	2.463	2.298	1.831	**1.585**
Lincoln	1.012	0.386	0.198	**0.195**	2.461	1.430	0.873	**0.866**
Coffee	1.013	0.470	0.268	**0.253**	2.473	2.681	2.454	**1.309**
C-3PO	1.013	0.344	**0.164**	0.199	2.474	1.314	**0.899**	0.923
Cheeseburger	1.014	0.283	**0.189**	0.208	2.466	1.561	1.160	**1.147**

ows. "Lincoln"[2] tests a Lambertian object in a complex lighting environment with multiple casted shadows. "Coffee"[3] involves a complex scene with a coffee mug and splashed liquid. "C3PO"[4] is a non-Lambertian object with a point light source. "Cheeseburger"[5] is a non-Lambertian, multiple albedo object with three-point lighting.

All models were rendered with Cycles renderer of Blender.[6] We added Gaussian noise with zero mean and standard deviation of 1.5 to the depth maps to simulate a depth sensor noise. The algorithm parameters were set to $\lambda_\rho = 0.1$, $\lambda_\beta^1 = 1$, $\lambda_\beta^2 = 1$, $\tau = 0.05$, $\sigma_c = \sqrt{0.05}$, $\sigma_d = \sqrt{50}$, $\lambda_z^1 = 0.004$, $\lambda_z^2 = 0.0075$, these values were carried throughout all our experiments. We evaluate the performance of each method by measuring the median of the depth error, and the 90th percentile of the depth error compared to the ground truth. The results are summarized in Table 3.1. An example of the accuracy improvement of our method can be easily seen in Fig. 3.7, which compares between the Thai Statue input errors and the output errors with respect to the ground truth.

We now show the qualitative results of the proposed framework from real data, captured by Intel's Real-Sense RGB-D sensor. First, we show how the proposed method handles texture, which usually leads to artifacts in shape-from-shading methods. In our experiment, we printed a text on a white page and captured it with our RGB-D scanner. Figure 3.8 shows how the texture copy artifact is mitigated by correctly modeling the scene albedos using λ_ρ.

Figure 3.9 compares between the reconstruction results of WZNSIT and the proposed framework in a real-world scenario of a mannequin captured under natural lighting. The proposed reconstruction procedure better captures the fine details.

Figure 3.10 illustrates how our algorithm handles real-world shapes with multiple albedos. The algorithm successfully reveals the letters and eagle on the shirt along with the "SF" logo, "champions" and even the stitches on the baseball cap. One

---

[2]http://3d.si.edu/downloads/27.

[3]http://www.blendswap.com/blends/view/56136.

[4]http://www.blendswap.com/blends/view/48372.

[5]http://www.blendswap.com/blends/view/68651.

[6]www.blender.org.

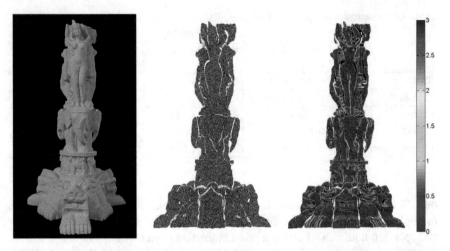

Fig. 3.7 Thai Statue error analysis. From left to right: Input color image. Error image of the raw depth map. Error image of the final result. Note how our diffuse shape refinement model reduces the initial surface errors

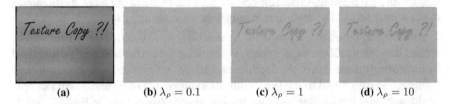

(a)   **(b)** $\lambda_\rho = 0.1$ **(c)** $\lambda_\rho = 1$ **(d)** $\lambda_\rho = 10$

Fig. 3.8 Texture copy. A correct albedo recovery model **b** mitigates texture copy artifact which the input figure (**a**) is prone to. The implications of poorly chosen albedo model can be easily seen in reconstructions (**c**) and (**d**). We note that $\lambda_\rho = 0.1$ was used throughout Sect. 3.6.1

should also notice that the algorithm was slightly confused by the grey "N" which is printed on the cap but do not stick out of it like the rest of the writing. We expect such results to be improved with stronger priors, however, incorporating such priors into real-time systems is beyond the scope of this paper.

Next, we show the robustness of our method to normal outliers. Such robustness is important for real-time performance, where segmenting the shape may cost precious time. In turn, background normals distort the shading recovery process, which might degrade the surface reconstruction. In this experiment, we run the method using the normals of the entire depth map. Thus, we deliberately distort the shading estimation process and examine how it affects the algorithm output. We ran this test on our method and on HLK. The results are presented in Fig. 3.11. We see that the proposed framework can gracefully handle a large amount of normal outliers, hence, it can be used in real-time scenarios with no need to separate the object from its background.

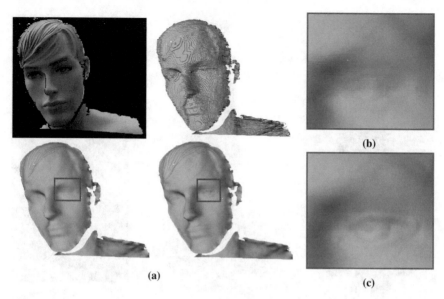

Fig. 3.9 Mannequin. a Upper Left: Color Image. Upper Right: Raw Depth. Bottom Left: Result of Wu et al. Bottom Right: Our diffuse shape refinement model Result. **b**, **c** Magnifications of the mannequin eye. The mannequin's hair and facial features can be easily recognized in our reconstruction

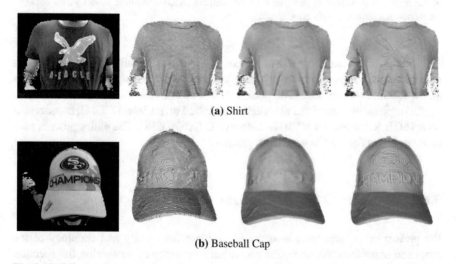

Fig. 3.10 Results of shape enhancement of real-world multiple albedo objects. Left to right: Color Image, Raw Depth, Bilateral Filtering, and the proposed diffuse shape refinement model. Note how surface wrinkles and small surface protrusions are now visible

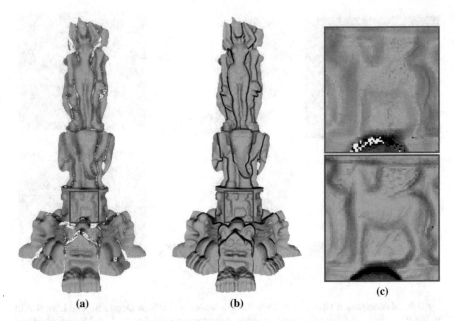

(a)   (b)   (c)

Fig. 3.11 Robustness to normal outliers: Left to right: HLK reconstruction with the entire depth map normals (**a**). Our method reconstruction with the entire depth map normals (**b**). Magnification of the results is presented in (**c**). The proposed diffuse shape refinement model yields accurate reconstruction despite the distorted shading

In Fig. 3.12, we can see how our method produces high quality reconstruction of a multiple albedo object without any prior knowledge of the shape or albedos. This is also a crucial aspect of real-time performance and everyday use in dynamic scenes.

An implementation of the algorithm was tested on an Intel i7 3.4 GHz processor with 16GB RAM and an NVIDIA GeForce GTX690 GPU. The entire process runs at about 25 fps for a 640×480 depth profiles.

3.6.2 Specular Shape Refinement

We performed several tests in order to evaluate the quality and accuracy of the proposed algorithm. We show the algorithm's accuracy in recovering the specular lighting of the scene and why it is vital to use an IR image instead of an RGB image. In addition, we demonstrate that the proposed framework is state-of-the-art, both visually and qualitatively.

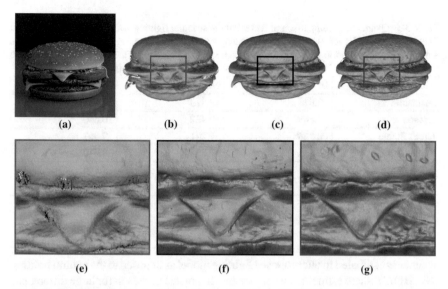

Fig. 3.12 **Handling a multiple albedo object. a** Color Image. **b** HLK Reconstruction. **c** WZNSIT Reconstruction. **d** Our diffuse shape refinement model Reconstruction. **e–g** Magnifications of HLK, WZNSIT, and Our Method, respectively. Note how the proposed framework sharply distinguish albedo changes

In order to test the specular lighting framework, we took 3D objects from the Stanford $3D$,[7] $123D$ Gallery[8] and Blendswap[9] repositories. For each model, we assigned a mix of diffuse and specular shaders and rendered them under an IR lighting scenario described in Sect. 3.4.2 (single light source) and natural lighting scenarios (multiple light sources) using the Cycles renderer in Blender.

To get a ground truth specularity map for each lighting scenario, we also captured each model without its specular shaders and subtracted the resulting images. We tested the accuracy of our model in recovering specularities for each lighting setup. We used Eqs. (3.18) and (3.21), to get the diffuse and ambient shading maps under IR lighting. For natural lighting, the diffuse and ambient shading were recovered using first and second-order spherical harmonics in order to have two models for comparison. In both lighting scenarios, the surface normals were calculated from the ground truth depth map.

The specular lighting is recovered using Eqs. (3.19) and (3.23), where the IR lighting direction $\vec{l}_p$ is calculated using the camera-projector calibration parameters. In the natural lighting scene, we use the relevant normalized coefficients of the first and second-order spherical harmonics in order to compute the general lighting direction.

---

[7] http://graphics.stanford.edu/data/3Dscanrep/.

[8] http://www.123dapp.com/Gallery/content/all.

[9] http://www.blendswap.com/.

Table 3.2 Quantitative comparison of RMSE of the specular lighting estimation in IR and natural lighting scenarios. IR refers to the lighting scenario described in Sect. 3.4.2, NL-SH1/2 represents a natural lighting scenario with first/second order spherical harmonics used to recover the diffuse and ambient shading, as well as $\vec{l}_p$. All values are in gray intensity units [0, 255]

Model	IR	NL-SH1	NL-SH2
Armadillo	**2.018**	12.813	11.631
Dragon	**3.569**	10.422	10.660
Greek Statue	**2.960**	7.241	9.067
Stone Lion	**4.428**	7.8294	8.640
Cheeseburger	**9.517**	17.881	19.346
Pumpkin	**10.006**	13.716	16.088

From the results in Table 3.2 we can infer that the specular irradiance can be accurately estimated in our proposed lighting model as opposed to the natural lighting (NL SH1/2) where estimation errors are much larger. The reason for large differences is that, as opposed to our lighting model, under natural illumination there are usually multiple light sources that cause specularities whose directions cannot be recovered accurately. An example of this can be seen in Fig. 3.13.

To measure the depth reconstruction accuracy of the proposed method we performed experiments using both synthetic and real data. In the first experiment, we used the $3D$ models with mixed diffuse and specular shaders and rendered their IR image and ground truth depth maps in Blender. We then quantized the ground truth depth map to 1.5mm units in order to simulate the noise of a depth sensor. We applied our method to the data and defined the reconstruction error as the absolute difference between the result and the ground truth depth maps. We compared our method's performance with the method proposed in [9, 50], as well as our diffuse refinement pipeline. The comparisons were performed in the specular regions of the objects according to the ground truth specularity maps. The results are shown in Table 3.3. A qualitative evaluation of the accuracy when the method is applied to the synthetic data is shown in Figs. 3.14 and 3.15.

In the second experiment, we tested our method under laboratory conditions using a structured-light $3D$ scanner to capture the depth of several objects. The camera-projector system was calibrated according to the method suggested in [80]. We reduced the number of projected patterns in order to obtain a noisy depth profile. To approximate an IR lighting scenario, we used a monochromatic projector and camera with dim ambient illumination. We also tested the algorithm with an Intel Real-Sense depth scanner, using the IR image and depth map as inputs. The camera-projector calibration parameters were acquired from the Real-Sense SDK platform. Although no accurate ground truth data was available for these experiments, we note that while all methods exhibit sufficient accuracy in diffuse areas, the proposed method is the only one that performs qualitatively well in highly specular areas as can be seen in Figs. 3.16 and 3.17.

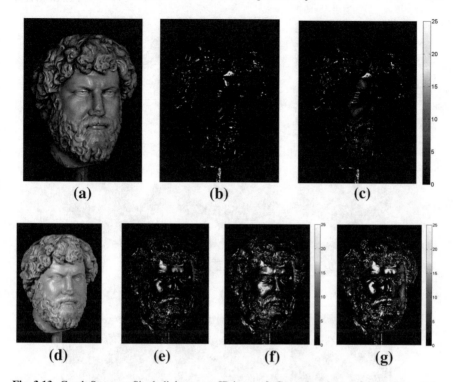

Fig. 3.13 Greek Statue: **a** Single light source IR image. **b** Ground truth specular irradiance map for (**a**). **c** Specular irradiance estimation error map. This is the absolute difference map between our predicted specular irradiance and the ground truth. **d** Multiple light source natural lighting (NL) image. **e** Specular lighting ground truth of (**d**). **f**, **g** Specular irradiance error maps of (**d**) as estimated using first (SH1) and second (SH2) order spherical harmonics respectively. Note the reduced errors when using a single known light source (**c**) as opposed to estimating multiple unknown light sources using spherical harmonics lighting models (**f**, **g**)

Table 3.3 Quantitative comparison of depth accuracy in specular areas. All values are in millimeters. Diffuse and specular are our diffuse framework and specular framework, respectively

Model	Median Error (mm)			90th % (mm)		
	Wu et al.	Diffuse	Specular	Wu et al.	Diffuse	Specular
Armadillo	0.335	0.318	**0.294**	1.005	0.821	**0.655**
Dragon	0.337	0.344	**0.324**	0.971	0.917	**0.870**
Greek Statue	0.306	0.281	**0.265**	0.988	0.806	**0.737**
Stone Lion	0.375	0.376	**0.355**	**0.874**	0.966	0.949
Cheeseburger	0.191	0.186	**0.168**	0.894	**0.756**	0.783
Pumpkin	0.299	0.272	**0.242**	0.942	0.700	**0.671**

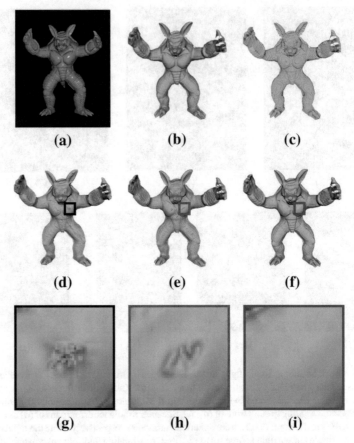

Fig. 3.14 Results for the simulated Armadillo scene, **a** Input IR image. **b** Ground truth model. **c** Initial Depth. **d–f** Results of Wu et al., the diffuse framework and the specular model respectively. **g–i** Magnifications of a specular area. Note how our surface is free from distortions in specular areas unlike the other methods

3.6.3 Learned Shape Refinement

The strength of the proposed method is demonstrated both qualitatively and quantitatively on 3D facial datasets and *in the wild* inputs We compare our results with the template-based method of [75], the 3DMM based method introduced as part of [82] and with the learning-based method of [71]. Unlike our method, all other methods require manual alignment as a preprocessing step. The state-of-the-art alignment framework of [83] is used to provide input for these algorithms.

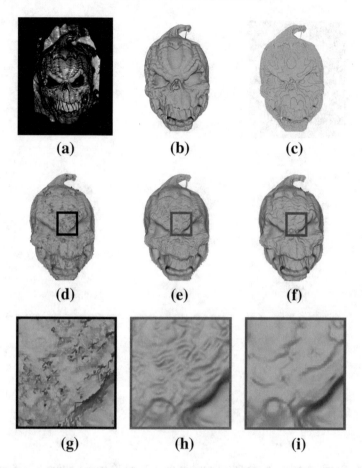

Fig. 3.15 Results for the simulated Pumpkin scene, **a** Input IR image. **b** Ground truth model. **c** Initial Depth. **d–f** Reconstructions of Wu et al., the diffuse framework and the specular model, respectively. **g–i** Magnifications of a specular area. Note the lack of hallucinated features in our method

In Fig. 3.18, we compare our reconstructions with a state-of-the-art method for reconstruction from multiple images [81]. The results show that our method is able to produce a comparable high quality geometry from only a single image. Figure 3.19 shows the robustness of our method to different poses. Finally, as described in Sect. 3.5.1, the FineNet framework can handle inputs with varying sizes. This is a vital property, as it allows the network to extract additional details as the quality of input image increases. Figure 3.20 shows how FineNet can gracefully scales up for 400×400 inputs although it was trained only on 200×200 images.

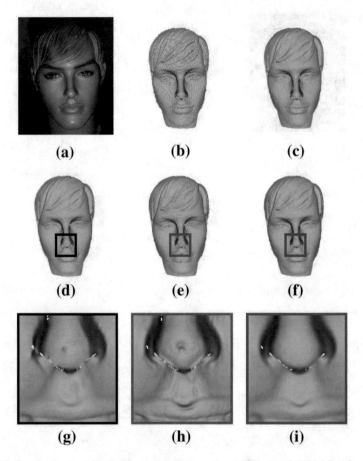

Fig. 3.16 Results for the lab conditions experiment, **a** Input IR image. **b** Initial Depth. **c** Result after bilateral smoothing. **d–f** Reconstructions of Wu et al., the diffuse framework and the specular model, respectively. **g–i** Magnifications of a specular region

We used the Face Recognition Grand Challenge dataset V2 [84] for a quantitative analysis of our results. This dataset has approximately two thousand color facial images with aligned ground truth depth. Each method provided an estimated depth image and a binary mask representing the valid pixels. For fair judgment, only pixels which were valid in all of the methods' outputs were used in the evaluation. Table 3.4 shows that our method produces the lowest depth error among the tested methods.

Finally, we show our qualitative results on 400×400 *in-the-wild* images of faces. As can be seen in Fig. 3.21, our method exposes the fine facial details as opposed to [71, 82] and is more robust to expressions and different poses than [75].

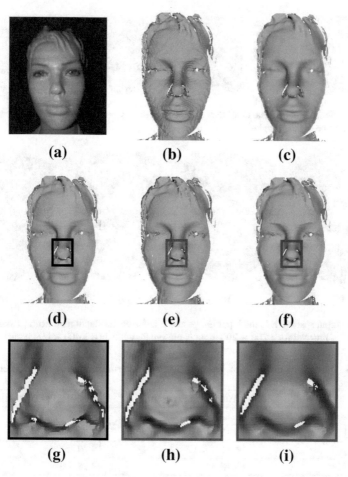

Fig. 3.17 Results from Intel's Real-Sense depth scanner, **a** Input IR image. **b** Initial Depth. **c** Result after bilateral smoothing. **d–f** Reconstructions of Wu et al., the diffuse framework and the specular model, respectively. **g–i** Magnifications of a specular region

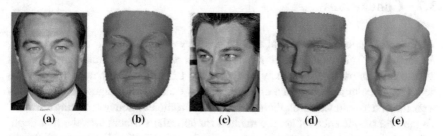

Fig. 3.18 **a** and **c** are two input images, **b** and **d** are their 3D reconstruction via the proposed CNN architecture. **e** is a reconstruction of the same subject, based on 100 different images recovered with the method proposed in [81]

Fig. 3.19 Method robustness. Our CNN architecture shows some robustness to extreme orientations, even in nearly 90° angles

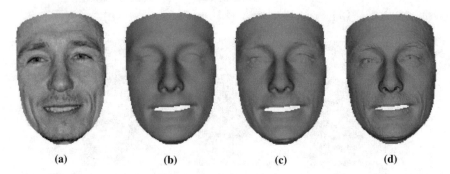

(a)   (b)   (c)   (d)

Fig. 3.20 Input scaling. **a** Is the input image and **b** is the coarse depth map from CoarseNet. In **c** the output of FineNet for a 200×200 input is presented, while in **d** a 400×400 input is used

Table 3.4 Quantitative comparison. We compare the depth estimation errors of the different methods vs. our CNN architecture

Method	Ave. Depth Err. (mm)	90% Depth Err. (mm)
Ours	**3.22**	**6.69**
Ref. [75]	3.33	7.02
Ref. [71]	4.11	8.70
Ref. [82]	3.46	7.36

3.7 Conclusions

We discussed two computational approaches, as well as a learning-based approach, for recovering surface details from a noisy depth map. The first framework handles diffuse objects using an intensity image, while the second framework refines the depth of specular objects based on shading cues from an IR image. These methods were the first to reconstruct explicitly the surface profiles without integrating normals. To the best of our knowledge, the method for specular surfaces was the first depth refinement framework that explicitly accounts for specularities. Furthermore, thanks to efficient optimization schemes both methods can run in real-time while achieving state-of-the-art accuracy. The FineNet CNN architecture can directly enhance depth details without any optimization process due to our efficient shape-from-shading based unsupervised training regime.

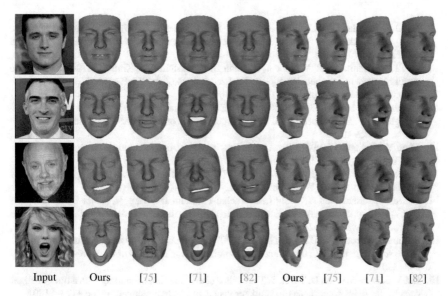

Fig. 3.21 Qualitative results. Input images are presented alongside the reconstruction results of different methods from two different viewpoints. Note that unlike the other methods, the proposed CNN architecture is robust to pose and expression variations, while still capturing subtle facial details

Acknowledgements We thank David Dovrat for his help with the implementation and Alon Zvirin for his help with the experiments. This research was partially supported by the Israel Ministry of Science and Technology grant number 3-14719 and the Technion Hiroshi Fujiwara Cyber Security Research Center and the Israel Cyber Directorate.

References

1. Horn BK (1970) Shape from shading: a method for obtaining the shape of a smooth opaque object from one view. PhD thesis, MIT
2. Bruckstein AM (1988) On shape from shading. Com Vis Graph Image Process 44(2):139–154
3. Kimmel R, Sethian JA (2001) Optimal algorithm for shape from shading and path planning. J Math Imaging Vis 14(3):237–244
4. Huang R, Smith WA (2011) Shape-from-shading under complex natural illumination. In: 18th IEEE international conference on image processing, 2011, pp 13–16
5. Johnson MK, Adelson EH (2011) Shape estimation in natural illumination. In: IEEE conference on computer vision and pattern recognition, 2011, pp 2553–2560
6. Zhang Q, Ye M, Yang R, Matsushita Y, Wilburn B, Yu H (2012) Edge-preserving photometric stereo via depth fusion. In: IEEE conference on computer vision and pattern recognition, pp 2472–2479
7. Yu LF, Yeung SK, Tai YW, Lin S (2013) Shading-based shape refinement of RGB-D images. In: IEEE conference on computer vision and pattern recognition (CVPR), pp 1415–1422
8. Haque S, Chatterjee A, Govindu VM (2014) High quality photometric reconstruction using a depth camera. In: IEEE conference on computer vision and pattern recognition (CVPR), pp 2283–2290

9. Or El R, Rosman G, Wetzler A, Kimmel R, Bruckstein AM (2015) RGBD-fusion: real-time high precision depth recovery. In: IEEE conference on computer vision and pattern recognition (CVPR), pp 5407–5416

10. Or-El R, Hershkovitz R, Wetzler A, Rosman G, Bruckstein AM, Kimmel R (2016) Real-time depth refinement for specular objects. In: Proceedings of the IEEE conference on computer vision and pattern recognition, pp 4378–4386

11. Richardson E, Sela M, Or-El R, Kimmel R (2017) Learning detailed face reconstruction from a single image. In: Proceedings of the IEEE conference on computer vision and pattern recognition, pp 1259–1268

12. Chen Y, Medioni G (1992) Object modelling by registration of multiple range images. Image Vis Comput 10(3):145–155

13. Digne J, Morel JM, Audfray N, Lartigue C (2010) High fidelity scan merging. In: Computer graphics forum, vol 29. Wiley Online Library, pp 1643–1651

14. Merrell P, Akbarzadeh A, Wang L, Mordohai P, Frahm JM, Yang R, Nistér D, Pollefeys M (2007) Real-time visibility-based fusion of depth maps. In: IEEE 11th international conference on, computer vision, pp 1–8

15. Schuon S, Theobalt C, Davis J, Thrun S (2009) Lidarboost: Depth superresolution for tof 3D shape scanning. In: IEEE conference on computer vision and pattern recognition (CVPR), pp 343–350

16. Cui Y, Schuon S, Chan D, Thrun S, Theobalt C (2010) 3D shape scanning with a time-of-flight camera. In: IEEE conference on computer vision and pattern recognition, pp 1173–1180

17. Tong J, Zhou J, Liu L, Pan Z, Yan H (2012) Scanning 3D full human bodies using kinects. IEEE Trans Visual Comput Graph 18(4):643–650

18. Newcombe RA, Davison AJ, Izadi S, Kohli P, Hilliges O, Shotton J, Molyneaux D, Hodges S, Kim D, Fitzgibbon A (2011) KinectFusion: real-time dense surface mapping and tracking. In: IEEE international symposium on Mixed and augmented reality, pp 127–136

19. Maier R, Kim K, Cremers D, Kautz J, Nießner M (2017) Intrinsic3d: high-quality 3d reconstruction by joint appearance and geometry optimization with spatially-varying lighting. In: Proceedings of the IEEE international conference on computer vision, pp 3114–3122

20. Zuo X, Wang S, Zheng J, Yang R (2017) Detailed surface geometry and albedo recovery from RGB-D video under natural illumination. In: Proceedings of the IEEE international conference on computer vision, pp 3133–3142

21. Sang L, Haefner B, Cremers D (2020) Inferring super-resolution depth from a moving lightsource enhanced RGB-D sensor: a variational approach. In: The IEEE winter conference on applications of computer vision, pp 1–10

22. Mac Aodha O, Campbell ND, Nair A, Brostow GJ (2012) Patch based synthesis for single depth image super-resolution. In: European conference on computer vision, 2012. Springer, pp 71–84

23. Hornáček M, Rhemann C, Gelautz M, Rother C (2013) Depth super resolution by rigid body self-similarity in 3D. In: IEEE conference on computer vision and pattern recognition, pp 1123–1130

24. Shechtman E, Irani M (2007) Matching local self-similarities across images and videos. In: IEEE conference on computer vision and pattern recognition, pp 1–8

25. Barnes C, Shechtman E, Finkelstein A, Goldman D (2009) PatchMatch: a randomized correspondence algorithm for structural image editing. ACM Transactions on Graphics-TOG 28(3):24

26. Li Y, Xue T, Sun L, Liu J (2012) Joint example-based depth map super-resolution. In: IEEE international conference on multimedia and expo, pp 152–157

27. Rosman G, Dubrovina A, Kimmel R (2012) Sparse modeling of shape from structured light. In: 2012 second international conference on 3D imaging, modeling, processing, visualization and transmission (3DIMPVT), IEEE, pp 456–463

28. Liu MY, Tuzel O, Taguchi Y (2013) Joint geodesic upsampling of depth images. In: IEEE conference on computer vision and pattern recognition, pp 169–176

29. Yang Q, Yang R, Davis J, Nistér D (2007) Spatial-depth super resolution for range images. In: IEEE conference on computer vision and pattern recognition, pp 1–8
30. Park J, Kim H, Tai YW, Brown MS, Kweon I (2011) High quality depth map upsampling for 3D-TOF cameras. In: IEEE international conference on computer vision, pp 1623–1630
31. Lee HS, Lee KM (2013) Simultaneous super-resolution of depth and images using a single camera. In: IEEE conference on computer vision and pattern recognition, pp 281–288
32. Lu S, Ren X, Liu F (2014) Depth enhancement via low-rank matrix completion, pp 3390–3397
33. Horn BK, Brooks MJ (1986) The variational approach to shape from shading. Comput Vis Graph Image Process 33(2):174–208
34. Kimmel R, Bruckstein AM (1995) Tracking level sets by level sets: a method for solving the shape from shading problem. Comput Vis Image Understand 62(1):47–58
35. Mecca R, Wetzler A, Kimmel R, Bruckstein AM (2013) Direct shape recovery from photometric stereo with shadows. In: 2013 international conference on 3DTV-conference, IEEE, pp 382–389
36. Mecca R, Tankus A, Wetzler A, Bruckstein AM (2014) A direct differential approach to photometric stereo with perspective viewing. SIAM J Imaging Sci 7(2):579–612
37. Zhang R, Tsai PS, Cryer JE, Shah M (1999) Shape-from-shading: a survey. IEEE Trans Pattern Anal Mach Intell 21(8):690–706
38. Durou JD, Falcone M, Sagona M (2008) Numerical methods for shape-from-shading: a new survey with benchmarks. Comput Vis Image Underst 109(1):22–43
39. Forsyth DA (2011) Variable-source shading analysis. Inter J Comput Vis 91(3):280–302
40. Quéau Y, Mélou J, Castan F, Cremers D, Durou JD (2017) A variational approach to shape-from-shading under natural illumination. In: International workshop on energy minimization methods in computer vision and pattern recognition, Springer, pp 342–357
41. Barron JT, Malik J (2015) Shape, illumination, and reflectance from shading. IEEE Trans Pattern Anal Mach Intell 1670–1687
42. Kar A, Tulsiani S, Carreira J, Malik J (2015) Category-specific object reconstruction from a single image. In: IEEE conference on computer vision and pattern recognition (CVPR), pp 1966–1974
43. Richter SR, Roth S (2015) Discriminative shape from shading in uncalibrated illumination. In: IEEE conference on computer vision and pattern recognition (CVPR), pp 1128–1136
44. Böhme M, Haker M, Martinetz T, Barth E (2010) Shading constraint improves accuracy of time-of-flight measurements. Comput Vis Image Underst 114(12):1329–1335
45. Han Y, Lee JY, Kweon IS (2013) High quality shape from a single RGB-D image under uncalibrated natural illumination. In: IEEE international conference on computer vision (ICCV), pp 1617–1624
46. Kadambi A, Taamazyan V, Shi B, Raskar R (2015) Polarized 3D: high-quality depth sensing with polarization cues. In: IEEE international conference on computer vision, pp 3370–3378
47. Choe G, Park J, Tai YW, So Kweon I (2014) Exploiting shading cues in kinect IR images for geometry refinement. In: IEEE conference on computer vision and pattern recognition (CVPR), pp 3922–3929
48. Chatterjee A, Govindu VM (2015) Photometric refinement of depth maps for multi-albedo objects. In: IEEE conference on computer vision and pattern recognition (CVPR), pp 933–941
49. Ti C, Yang R, Davis J, Pan Z (2015) Simultaneous time-of-flight sensing and photometric stereo with a single tof sensor. In: IEEE conference on computer vision and pattern recognition (CVPR), pp 4334–4342
50. Wu C, Zollhöfer M, Nießner M, Stamminger M, Izadi S, Theobalt C (2014) Real-time shading-based refinement for consumer depth cameras. In: ACM transactions on graphics (Proceedings of SIGGRAPH Asia 2014), vol 33
51. Haefner B, Quéau Y, Möllenhoff T, Cremers D (2018) Fight ill-posedness with ill-posedness: single-shot variational depth super-resolution from shading. In: Proceedings of the IEEE conference on computer vision and pattern recognition, pp 164–174
52. Peng S, Haefner B, Quéau Y, Cremers D (2017) Depth super-resolution meets uncalibrated photometric stereo. In: Proceedings of the IEEE international conference on computer vision workshops, pp 2961–2968

53. Quéau Y, Durou JD, Aujol JF (2018) Normal integration: a survey. J Math Imaging Vis 60(4):576–593
54. Langguth F, Sunkavalli K, Hadap S, Goesele M (2016) Shading-aware multi-view stereo. In: European conference on computer vision, Springer, pp 469–485
55. Wu S, Huang H, Portenier T, Sela M, Cohen-Or D, Kimmel R, Zwicker M (2018) Specular-to-diffuse translation for multi-view reconstruction. In: Proceedings of the European conference on computer vision (ECCV), pp 183–200
56. Guo K, Xu F, Yu T, Liu X, Dai Q, Liu Y (2017) Real-time geometry, albedo, and motion reconstruction using a single rgb-d camera. ACM Trans Graph (TOG) 36(3):32
57. Quéau Y, Mélou J, Durou JD, Cremers D (2017) Dense multi-view 3d-reconstruction without dense correspondences. arXiv:1704.00337
58. Liu-Yin Q, Yu R, Agapito L, Fitzgibbon A, Russell C (2017) Better together: joint reasoning for non-rigid 3d reconstruction with specularities and shading. arXiv:1708.01654
59. Grosse R, Johnson MK, Adelson EH, Freeman WT (2009) Ground-truth dataset and baseline evaluations for intrinsic image algorithms. In: International conference on computer vision, pp 2335–2342
60. Basri R, Jacobs DW (2003) Lambertian reflectance and linear subspaces. IEEE Trans Pattern Anal Mach Intell 25(2):218–233
61. Ramamoorthi R, Hanrahan P (2001) An efficient representation for irradiance environment maps. In: Proceedings of the 28th annual conference on computer graphics and interactive techniques, ACM, pp 497–500
62. Land EH, McCann JJ (1971) Lightness and retinex theory. J Opt Soc Amer 61(1):1–11 Jan
63. Barron JT, Malik J (2012) Shape, albedo, and illumination from a single image of an unknown object. In: Vision computer, recognition pattern, IEEE Computer Society, Washington, DC, USA, pp 334–341
64. Chang J, Cabezas R, Fisher III JW (2014) Bayesian nonparametric intrinsic image decomposition. In: European conference on computer vision 2014. Springer, pp 704–719
65. Wu C, Tai XC (2010) Augmented lagrangian method, dual methods, and split Bregman iteration for ROF, vectorial TV, and high order models. SIAM J Img Sci 3:300–339
66. Sochen N, Kimmel R, Malladi R (1998) A general framework for low level vision. IEEE Trans Image Process 7(3):310–318
67. Roussos A, Maragos P (2010) Tensor-based image diffusions derived from generalizations of the total variation and Beltrami functionals. In: IEEE international conference on image processing (ICIP), IEEE, pp 4141–4144
68. Wetzler A, Kimmel R (2011) Efficient Beltrami flow in patch-space. In: Scale space and variational methods in computer vision (SSVM), pp 134–143
69. Rosman G, Bronstein AM, Bronstein MM, Tai XC, Kimmel R (2012) Group-valued regularization for analysis of articulated motion. In: NORDIA workshop, European conference on computer vision (ECCV), Springer, pp 52–62
70. Ping-Sing T, Shah M (1994) Shape from shading using linear approximation. Image Vis Comput 12(8):487–498
71. Richardson E, Sela M, Kimmel R (2016) 3D face reconstruction by learning from synthetic data. In: 2016 international conference on 3D vision (3DV), IEEE, pp 460–469
72. Blanz V, Vetter T (1999) A morphable model for the synthesis of 3D faces. In: Proceedings of the 26th annual conference on computer graphics and interactive techniques, ACM Press/Addison-Wesley Publishing Co., pp 187–194
73. Hariharan B, Arbeláez P, Girshick R, Malik J (2015) Hypercolumns for object segmentation and fine-grained localization. In: Proceedings of the IEEE conference on computer vision and pattern recognition, pp 447–456
74. Parkhi OM, Vedaldi A, Zisserman A (2015) Deep face recognition. British Mach Vis Conf 41(1–41):12
75. Kemelmacher-Shlizerman I, Basri R (2011) 3D face reconstruction from a single image using a single reference face shape. IEEE Trans Pattern Anal Mach Intell 33(2):394–405

76. Yoon Y, Choe G, Kim N, Lee JY, Kweon IS (2016) Fine-scale surface normal estimation using a single NIR image. In: European conference on computer vision, pp 486–500
77. Bansal A, Russell B, Gupta A (2016) Marr revisited: 2D-3D alignment via surface normal prediction. In: The IEEE conference on computer vision and pattern recognition (CVPR), pp 5965–5974
78. Sengupta S, Kanazawa A, Castillo CD, Jacobs DW (2018) Sfsnet: learning shape, reflectance and illuminance of faces in the wild'. In: Proceedings of the IEEE conference on computer vision and pattern recognition, pp 6296–6305
79. Curless B, Levoy M (1996) A volumetric method for building complex models from range images. In: Proceedings of the ACM conference on computer graphics and interactive techniques, SIGGRAPH, pp 303–312
80. Zhang S, Huang PS (2006) Novel method for structured light system calibration. Opt Eng 45(8):083601–1–083601–8
81. Roth J, Tong Y, Liu X (2016) Adaptive 3D face reconstruction from unconstrained photo collections, CVPR
82. Zhu X, Lei Z, Yan J, Yi D, Li SZ (2015) High-fidelity pose and expression normalization for face recognition in the wild. In: Proceedings of the IEEE conference on computer vision and pattern recognition, pp 787–796
83. Kazemi V, Sullivan J (2014) One millisecond face alignment with an ensemble of regression trees. In: The IEEE conference on computer vision and pattern recognition (CVPR)
84. Phillips PJ, Flynn PJ, Scruggs T, Bowyer KW, Chang J, Hoffman K, Marques J, Min J, Worek W (2005) Overview of the face recognition grand challenge. In: 2005 IEEE computer society conference on computer vision and pattern recognition (CVPR'05), vol 1. IEEE, pp 947–954

Chapter 4
Non-Rigid Structure-from-Motion and Shading

Mathias Gallardo, Toby Collins, and Adrien Bartoli

Abstract We show how photometric and motion-based approaches can be combined to reconstruct the 3D shape of deformable objects from monocular images. We start by motivating the problem using real-world applications. We give a comprehensive overview of the state-of-the-art approaches and discuss their limitations for practical use in these applications. We then introduce the problem of Non-Rigid Structure-from-Motion and Shading (NRSfMS), where photometric and geometric information is used for reconstruction, without prior knowledge about the shape of the deformable object. We present in detail the first technical solution to NRSfMS and close the chapter with the main remaining open problems.

4.1 Introduction

Deformable 3D reconstruction aims to recover the 3D shape of deformable objects from one or more 2D images. While 3D reconstruction of rigid objects is well understood with robust methods and commercial products [1, 2], deformable 3D reconstruction is still an open challenge. Taking up these challenges is important because many objects of interest are deformable, including faces, bodies, organs, clothes, and fabrics. Furthermore, 2D cameras are by far the most common types of imaging sensors in use today, yielding a broad range of useful applications for passive methods, as discussed in the next section.

M. Gallardo (✉)
EnCoV, Institut Pascal, UMR 6602, CNRS/UBP/SIGMA, 63000 Clermont-Ferrand, France
e-mail: Mathias.Gallardo@gmail.com

T. Collins
IRCAD and IHU-Strasbourg, 1 Place de l'Hôpital, 67000 Strasbourg, France
e-mail: toby.collins@ircad.fr

A. Bartoli
EnCoV, Institut Pascal, UMR 6602, CNRS/UBP/SIGMA, 63000 Clermont-Ferrand, France
e-mail: Adrien.Bartoli@gmail.com

© Springer Nature Switzerland AG 2020   115
J.-D. Durou et al. (eds.), *Advances in Photometric 3D-Reconstruction*,
Advances in Computer Vision and Pattern Recognition,
https://doi.org/10.1007/978-3-030-51866-0_4

Along with passive methods which only use 2D images, active methods with depth sensors have also tackled this problem. This has been done for instance using stereo-cameras [79], infrared projectors or time-of-light systems with Kinect [40, 55] and more elaborated systems such as color photometric stereo techniques [12, 31]. Despite their impressive results, active depth sensors suffer from inherent limitations: some have a restricted range (they cannot sense depth when the object is too far from or too close to the sensors), others have a significantly higher power consumption than RGB cameras, and some others are often strongly affected by outdoor illumination conditions. There may also be physical restrictions, such as in endoscopic applications, where it is not possible to use bulky active vision sensors. Finally, there are billions of monocular cameras used every day on mobile devices, which yields a huge potential for usage and commercialization and underlines the need for solving the problem of monocular deformable 3D reconstruction.

Four main paradigms have emerged to tackle deformable reconstruction with monocular cameras: Shape-from-Template (SfT), Non-Rigid Structure-from-Motion (NRSfM), Shape-from-Shading (SfS) and neural network-based reconstruction (NNR). We now briefly summarize them. SfT uses a known *template* of the object and at least one image of the object being deformed. It works by registering the object to the input image and deforming the template of the object accordingly in 3D [6, 71]. NRSfM uses multiple monocular images and recovers the 3D shape of the deforming object in each image [11]. This paradigm is much harder to solve than SfT because no template is available and consequently, the physical structure of the object is unknown a priori. Figure 4.2 illustrates both paradigms. SfS only uses a single image and recovers the depth or surface normal at each pixel. SfS works exclusively with shading information which links surface geometry, surface reflectance, scene illumination, pixel intensity, and camera response function [37]. SfS is often very difficult to use in practice, mainly because it is generally ill-posed, requires a complete *photometric calibration* of the scene *a priori* and suffers discrete convex/concave ambiguities. NNR approaches predict the 3D shape of an object or the depth-map of a scene from a single image using a trained neural network. Most of these methods pose the problem as a supervised learning task. It works well for common object classes with very large datasets available, such as man-made objects [65] and faces [67] using specific low-dimensional deformation models. This category has not shown to enable 3D reconstruction of objects under very high dimensional deformations, which notably limits its applicability. Two other shortcomings are the need for large amounts of annotated training data, comprising images and known 3D deformation pairs, which are hard to obtain with real data, and the need for a training phase which may not be practical in several real applications, when the template is acquired at run-time. For these reasons, the most practical paradigms are currently SfT and NRSfM. Nevertheless, neural networks can be used in conjunction with SfT and NRSfM to provide state-of-the-art solutions to intermediate problems, such as motion estimation and feature extraction.

Most SfT and NRSfM methods use the apparent motion of the object's surface, also called motion cue. That is, by knowing the relative movement between the surface template and the image, or between images, they infer the 3D deformation.

However, the motion cue is often insufficient to infer the 3D shape of a deforming object, because motion can be explained by possibly infinitely many 3D deformations (the so-called depth ambiguity). To fix this, SfT and NRSfM methods, as with other deformable 3D reconstruction approaches, use deformation priors, which we discuss in detail in Sect. 4.3.1.1. Despite the inclusion of deformation priors, SfT and NRSfM methods generally fail in two cases: when the object is poorly-textured or when it deforms non-smoothly. Figures 4.6 and 4.7 illustrate this with some reconstructions from state-of-the-art NRSfM methods [14, 60]. At poorly-textured regions, motion information is sparse and accurate reconstruction becomes difficult. In the last years, to overcome these limitations, some methods proposed to complement the motion cue with the shading cue. Unlike motion, shading can be used to reconstruct textureless surfaces, as it is considered the most important visual cue for inferring shape details at textureless regions [61].

Contributions
We propose to combine shading with NRSfM in order to reconstruct densely-textured and poorly-textured surfaces under non-smooth deformations. We refer to this problem as NRSfMS (Non-Rigid Structure-from-Motion and Shading). We are specifically interested in solving this problem for objects with unknown spatially-varying albedo, which is the situation in most practical cases. However, we must know albedos in order to apply shading constraints. Therefore, our problem is to simultaneously and densely estimate non-rigid 3D deformation from each image together with spatially-varying surface albedo. We assume deformation is either piecewise or globally smooth, which allows us to reconstruct creasing or wrinkled surfaces. We assume that the albedo is piecewise smooth, which is very common for man-made objects. Furthermore, this assumption on albedo reduces the potential ambiguity between smooth albedo changes and smooth surface orientation changes. This problem has not been tackled before. It is a crucial missing component for densely reconstructing images in unconstrained settings and may then enlarge the spectrum of deformations and surfaces for more real-world applications.

4.2   Applications of Monocular Deformable 3D Reconstruction

Research on monocular deformable 3D reconstruction has raised considerable interest in its applications in many domains including medical image processing, special effects, data-driven simulation, Augmented Reality (AR) games, and soft body mechanics. Some examples are shown in Fig. 4.1.

There are many important applications in medical AR with Minimally Invasive Surgery (MIS) and more precisely with laparoscopic surgery. This advanced surgery technique is performed by inserting, through small incisions, small surgical instruments and a laparoscope, which is a thin, tube-like instrument with a light and a digital camera. The surgeon uses the live video from the camera to perform the surgery. This

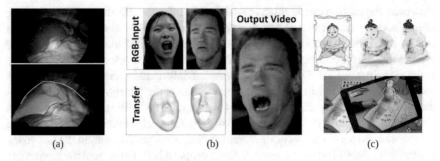

Fig. 4.1 Applications of monocular deformable 3D reconstruction: **a** medical imaging [44], **b** post-production movie editing [77] and **c** AR gaming [50]

reduces the patient's trauma and shortens the recovery time, compared to open surgeries. However, during MIS, surgeons face three main problems: the viewpoint is limited, the localization in 3D and the perception of depth become harder, and they cannot see the locations of important subsurface structures such as a tumor or major vessels. AR appears to be a very suitable way to give a real-time feedback during MIS. This is done by augmenting the live video with a deformable model of the organ, including its surface and internal structures. The deformable model can be constructed from a preoperative medical image, such as MRI or CT, and the task of registering the organ model to the laparoscopic video is an SfT problem. Using a monocular laparoscope, a deformable registration of a preoperative template of a liver (obtained from CT) was presented in [44]. This permits one to register at the same time the tumor (in green) and the internal structures of the liver such as veins (in blue).

Another application area is video post-production. Video editors are often required to modify videos after the recording, by removing, introducing or modifying content. When the content is deformable, this can be highly labor intensive. Most videos are not recorded with depth sensors, which makes monocular methods extremely valuable. A real-time technique of facial performance capture and editing on 2D videos was proposed in [77]. It works by reconstructing the 3D faces of a source actor and of a target actor, and transfer the facial expression of the source actor to the target actor.

Another large application domain is AR gaming. The idea is to offer players new gameplay experiences and a different game environment that combine virtual content with the real-world environment. Nearly all AR games assume the scene to be rigid. Recently new games have been presented, enabled by SfT. For instance, an AR coloring book application is presented in [50]. The idea is for a player to interactively color a virtual 3D character by coloring a 2D sketch of the character printed in a book, using color pencils. An SfT algorithm is used to register a template of each book page and estimate the deformation of the visible page. This allows registration of the colored page with the virtual character, which then allows the transfer of pencil colors to the virtual character and visualization in real-time.

4.3   Background on Deformable Monocular 3D Reconstruction and Shading

Since the first works on deformable monocular 3D reconstruction [11, 32], many technical and theoretical aspects have been explored in both NRSfM and SfT. The main ones are *(i)* shape and deformation modeling, *(ii)* data constraints extracted from the input images, *(iii)* 3D shape inference and *(iv)* the use of temporal coherence. As our main contribution relates to NRSfM, we thoroughly review NRSfM and each of the above directions. We then propose an overview of 3D reconstruction using shading and especially focus on some works which integrate shading with SfT. This is motivated by the fact that SfT and NRSfM are closely related.

4.3.1   Non-Rigid Structure-from-Motion

The goal of NRSfM is to recover a deformable object's shape from a set of unorganized images or a video, as depicted in Fig. 4.2.

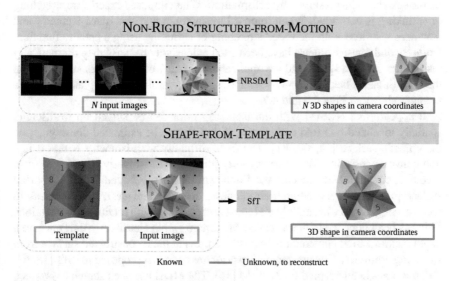

Fig. 4.2 Illustration of the problems of NRSfM and SfT

4.3.1.1 Deformation Priors in NRSfM

Deformation priors are required to make NSRfM well-posed. Three classes of deformation priors have emerged: *statistical* priors, *physics-based* priors and *temporal smoothness* priors.

Statistical priors have been formulated in two main ways: low-rank shape bases and low-rank trajectory bases. Both ways use a reduced space of modes to model the shapes, which are learned during the reconstruction, i.e., which is thus the joint estimation of the model modes, weights, and the camera poses. These modes are usually constrained to lie in a linear space spanned by a small number of an unknown 3D shape bases [11], or of unknown 3D trajectory bases [4, 18]. Both approaches reduce the problem dimensionality, however, they present three limitations. They tend to require a large number of images and short-baseline data to achieve good results, and they lose the ability to model high-frequency deformations, i.e., discontinuities, such as creases.

Physics-based deformation priors operate very differently to statistical models, and restrict the space of possible deformations according to physical properties of the object's material. The most common *physics-based* priors is isometry or quasi-isometry [14, 15, 60, 81, 83, 85]. It enforces the geodesic distance between two points on the surface to be preserved by deformation. When imposed exactly, no stretching or shrinking is permitted. When imposed inexactly, it is called quasi-isometry and penalizes solutions with increased stretching or shrinking using a penalty function. Isometry and quasi-isometry have been used extensively because they dramatically restrict the solution space, and are applicable for many object classes such as those made of thick rubber, tightly-woven fabrics, paper, cardboard, and plastics such as the ones shown in Figs. 4.6 and 4.7.

It appears that NRSfM with the isometric prior can be solved up to discrete, spatially localized two-fold ambiguities if motion can be estimated densely across the object's surface [14, 60, 81]. The main difficulty with isometry is that it is a non-convex constraint. The *inextensibility* prior relaxes isometry in order to form a convex constraint. It prevents the Euclidean distance between two neighboring surface points from exceeding their geodesic distance. However, it is too weak to reconstruct geometry accurately and must be combined with additional constraints. This has been done using the so-called Maximum Depth Heuristic (MDH), where a depth maximization constraint is imposed to prevent the reconstructed surface from shrinking arbitrarily. The inextensibility prior has been first proposed for SfT [13, 63, 70], but it has been adapted for NRSfM [16]. The MDH has been shown to produce very good reconstructions when the perspective effects of the camera are strong.

Temporal smoothness priors assume that the object deforms smoothly over time. These priors have been mainly used through two approaches: *(i)* using temporal smoothing constraint [4, 29] and *(ii)* initializing the shape of an input image using the one of the previous input image [85]. One important advantage of the approach *(ii)* is that the problem is more constrained. This may then provide more accurate reconstructions since it optimizes an initial solution. However, this can turn as a shortcoming. The solution can be stuck in a local minimum if the solution from the previous frame is wrong, because of tracking loss which may happen in case of sudden illumination changes or occlusions. Temporal smoothness can also be used to assist correspondence estimation generally using optical flow approaches to obtain dense correspondences [29, 85].

4.3.1.2   Data Constraints in NRSfM

NRSfM methods rely fundamentally on motion constraints, and these can be divided into two types: *correspondence constraints*, which assume the correspondences are computed a priori [14, 15, 29, 30, 60, 81, 83], and *direct constraints* those which compute correspondence jointly to deformation inference using brightness constancy [85]. By far, the most common constraints are motion constraints, however, *contour constraints* have been also used in NRSfM [85].

Correspondence constraints force 3D points on the surface to project at their corresponding 2D points in each input image. The points used by these constraints are usually obtained by matching features [49], or tracking points [29, 74], in the images. Correspondence constraints have four main limitations. First, feature-based matching methods may fail to establish correspondences without errors. Second, the computational time to extract features, compute descriptors, and perform the matching can be long without high performance GPUs. Third, tracking-based methods require short-baseline input images. Fourth, they work well only for densely-textured objects with discriminative texture, which are not common in most real practical applications, particularly with man-made objects and many natural objects.

Direct constraints work by maximizing the photometric agreement, i.e., brightness constancy, between the input images [85]. Their main advantage is to provide denser motion constraints than correspondence constraints. They have, however, three main limitations. First, they are highly non-convex and they require iterative optimization. Because of non-convexity, they are usually applied in a frame-to-frame tracking setup. Second, direct constraints are sensitive to strong photometric changes which may be induced by complex deformations or complex illuminations. Third, direct constraints may require reasoning about surface visibility (they should be deactivated at surface regions that are occluded).

Contour constraints force the object's occluding contours to align with the corresponding contours in the images [85]. Contour constraints are interesting because they do not depend on the surface's texture, and therefore, are applicable for poorly-textured and even non-textured surfaces. There exist two types of contour constraints: silhouette contour constraints and boundary contour constraints. Silhouette contour constraints work by forcing the object's silhouette to align with silhouette contours detected in the input image and can be used for surfaces and volumes. These constraints have not been used in NRSfM yet, mainly because they are very difficult to use without any prior on the object's shape. Boundary contour constraints are applicable for open surface templates such as a piece of paper. They work by enforcing the boundary contour projects to image edges [85]. Similarly, to direct constraints, contour constraints are highly non-convex, usually enforced iteratively, and require a good initial estimate. However, they are also difficult to apply robustly, particularly with strong background clutter.

4.3.1.3   Local and Global Methods to NRSfM

Another important way to characterize NRSfM methods is whether they reconstruct a surface using local surface regions (usually called local methods), or whether they reconstruct the whole surface at once (usually called global methods). *Local methods* work by dividing the surface into local regions, reconstructing each region individually, and then reconstructing the full surface using surface continuity. Most local methods assume isometric deformations. They mainly differ by the way they locally model the surface: piecewise planes [76, 81], quadrics [23], or Partial Differential Equations (PDEs) [14, 60]. Their advantages are that they can be fast and can provide closed-form solutions. However, they also produce sub-optimal results, because they do not enforce the physical prior globally over the whole surface, and they may be unstable and present ambiguities. *Global methods* use instead constraints acting over the whole surface. These methods produce large, non-convex optimization problems that cannot lead to closed-form solutions. They generally use energy minimization frameworks for optimization. This allows them to handle more complex deformations and to use more complex constraints, leading to potentially more accurate reconstructions. However, they generally require high computation time and a good initial solution, and they are often difficult to optimize and not easily parallelizable.

4.3.1.4   Unorganized Image Sets Versus Video Inputs

A final way to categorize NRSfM methods is if they operate with unorganized image sets [14, 60, 81] or videos [4, 16, 23, 29, 76, 83] as inputs. A fundamental difference between these two settings is that temporal continuity can be exploited in the latter setting but not in the former setting. Typically for video inputs, methods work in an incremental style where new frames are added to the optimization process while fixing unknowns in past frames [4, 23, 29]. This strategy is used to manage the growing number of unknowns with video inputs, allowing it to scale well for long videos.

4.3.2   3D Reconstruction Using Shading

Shading relies on the photometric relationship between surface geometry, surface reflectance, illumination, the camera response, and pixel intensity. This relationship, also called the shading equation, provides one constraint on the surface normal at any given pixel. Shading is a powerful visual cue because, unlike motion, it can constrain

3D shape at poorly-textured surface regions and recover complex deformations in such surfaces [62]. Shading has been first used alone in the paradigm of SfS and then in other 3D reconstruction problems and in SfT.

4.3.2.1 Shape-from-Shading

SfS consists in using shading to recover the 3D shape of an object from a single image. Precisely, it recovers the surface normal at each pixel of the image. SfS has been intensively studied in the last decades and the SfS literature can be explored through four main components: *(a)* the camera projection model, *(b)* the illumination model, *(c)* the surface reflectance model, and *(d)* the 3D shape inference algorithm. For *(a)*, SfS has been first studied with the orthographic camera [37, 61] and then the perspective camera model [64, 75]. For *(b)*, most of the existing methods assume a distant light source, but more complex illumination models are also used, such as the near-point lighting with fall-off [58, 64]. Most SfS methods also assume known illumination. For *(c)*, a very common assumption of reflectance is the Lambertian model [21, 37, 39, 43, 47, 58, 61, 64, 68, 69, 78, 88]. The Lambertian model assumes a diffuse reflection of the surface, ı.e., the surface luminance is independent from the viewing angle. Non-Lambertian reflectance models are also studied [3, 46], such as the Oren-Nayar and Ward models which, respectively, take into account the micro-facets reflections and specular reflections. Nearly all methods assume either constant and fixed albedo or known albedo. This is because SfS is fundamentally an ill-posed problem with unknown varying albedo. Some works, however, propose solutions to handle multi-albedo surfaces. To handle multiple albedos, [5] forms a complex energy function which simultaneously solves several problems related to the photometric formation of the image, namely SfS, intrinsic images decomposition, color constancy and illumination estimation. For *(d)*, SfS methods can be divided in six subcategories: *(i)* propagation approaches [64], *(ii)* local approaches [61], *(iii)* linear approaches [78], *(iv)* convex minimization approaches [21], *(v)* non-convex minimization approaches [5], and *(vi)* learning-based approaches [68].

Despite the great interest drawn by SfS and the diversity of the proposed approaches, almost all of the existing SfS methods share the same shortcomings. First, they assume the albedo values and the scene illumination to be known, ı.e., they require a complete photometric calibration, as the survey [20] shows. Second, they suffer from convex/concave ambiguities [9, 38]. Third, they cannot handle depth discontinuities and provide a surface solution up to a global scale factor.

4.3.2.2 Extending SfS to Multiple Images

Shading has been used previously in several other 3D reconstruction problems. These include photometric stereo [12, 31, 91], multi-view SfS [41, 86], multi-view reconstruction [8, 42, 45, 79, 87], SfM and SfS [42]. Their main limitations are that they work for rigid objects or/and use impractical setups.

Photometric stereo is the extension of SfS using multiple light sources. The images taken under different illumination contain no motion. This is one big difference between photometric stereo and the other extensions of SfS. It has shown great success for reconstructing high-accuracy surface details with unknown albedo such as [12, 31, 91]. However, it requires a special hardware setup where the scene is illuminated by a sequence of lights placed at different points in the scene. This setup is not applicable in many situations. Another limitation is that the scene is assumed to be rigid during the acquisition [12, 91]. [31] proposes, however, a photometric stereo technique which works for deforming surfaces.

Multi-image SfS methods, such as [41, 86], have shown that using shading and a collection of images, from monocular [41] or several tracked cameras [86], provide reasonably good reconstructions of poorly-textured surfaces such as statues or bones. Multi-view reconstruction methods, such as [8, 45, 79, 87], have shown that shading reveals fine details for e.g., clothes or faces. However, these methods assume rigid objects, use two or more cameras and require a special design of the scene, which may not be practical.

Shading has also been used in rigid SfM [42], which uses multiple images showing a rigid object. This approach initializes the surface using motion through SfM and MVS and then refines it by combining motion with shading information. Unlike the other extensions of SfS, this approach requires to solve a registration problem, to link pixel information across different images. One limitation may come from the difficulty of establishing correspondences accurately. However, because of the MVS constraints, this approach may achieve higher accuracy than photometric stereo at both textured and textureless regions.

4.3.2.3   Existing Methods to Solve SfT with Shading

We briefly present the principle of SfT regarding the directions *(i)* shape and deformation model, *(ii)* image data constraint, and *(iii)* 3D shape inference. We then describe how shading has been used in SfT. We refer the readers to [25], for more details on SfT.

Shape-from-Template
The goal of SfT is to register and reconstruct the 3D shape of a deforming object from a single input image, using a template of the object. This is illustrated in Fig. 4.2. The template is a textured 3D model of the surface in a rest position and the problem is solved by determining the 3D deformation that transforms the template into camera coordinates. The main difference between SfT and NRSfM is that in NRSfM the object's template is not provided a priori, and this makes it a considerably harder problem.

The template brings strong object-specific prior knowledge to the problem. It comprises a *shape model*, an *appearance model* and a *deformation model*. The *shape* model represents the object's 3D shape in a fixed reference position. The *appearance* model is used to describe the photometric appearance of the object. The *deformation* model is used to define the transformation of the template's reference shape and the space of possible deformations. For this, most methods use dimensionality reduction through *smooth parameterizations*, *explicit smoothing* priors or *physics-based* priors. *Smooth parameterizations* have included thin-plate splines and B-splines, and reduce dimensionality by modeling deformation with a discrete set of control points [56, 71, 73]. Smooth parameterizations reduce in general the cost of optimization, however, they lose the ability to model high-frequency deformations, i.e., discontinuities, such as creases. *Explicit smoothing* priors penalize non-smooth deformations explicitly. They use a smoothing term within an energy-based optimization, usually using the ℓ_2 norm [7, 13]. This norm strongly penalizes non-smooth deformations and provides strong problem regularization, but can prevent the formation of discontinuities such as creases. *Physics-based* priors in SfT work in a very similar manner than in NRSfM. The most commonly used is the *isometry* prior [6, 15, 48, 71], however, other priors have been studied: inextensibility [13, 63, 70], conformal (angle preservation) [6, 51] and elasticity [35, 53, 59].

Data constraints must be extracted from the input image in order to match the template's shape with the object's true shape. Similar to NRSfM, the most common data constraints are motion constraints [6, 13, 17, 52, 56, 57, 71, 89], but some methods also rely on contour [72, 84] and shading constraints.

Combining Motion and Shading in SfT
Shading has been also used as a complementary visual cue in SfT [27, 48, 51, 54, 82]. These methods differ in the way the problem is modeled and optimized. [54, 82] use motion and shading information sequentially, and not in an integrated manner. We refer to these as *non-integrated* approaches. The proposed approaches are difficult to use in practice because of several significant drawbacks. [82] requires a full photometric calibration a priori and the illumination to be the same at training and test time. [54] requires to know the reflectance of the template and only works for smooth deformations.

By contrast, in [27, 48, 51], shading, motion, and deformation priors are integrated together into a single non-convex energy function which is minimized through iterative refinement. We refer to these as *integrated* approaches. Their advantage is that they combine constraints from multiple cues simultaneously, to improve reconstruction accuracy, which is not possible with *non-integrated* approaches. [48, 51] simplify the problem by assuming a rigid observation video is available prior to deformable reconstruction. This is a video of the object taken from different viewpoints before any deformation occurs and is used for reconstructing the template. Reconstruction then proceeds with an SfT-based approach. The main limitation of this is that it requires control of the environment to ensure the object does not deform during the observation video. This is not often possible in real applications. Furthermore, such methods assume the scene illumination is constant and fixed during the rigid observation video, which is not always possible to achieve.

Current Limitations
As Figs. 4.6 and 4.7 illustrate, poorly-textured surfaces present an important limitations of nearly all state-of-the-art NRSfM and SfT methods, particularly with creases.

The reason is that motion information, which is the most used data constraint, is fundamentally insufficient to reconstruct textureless surface regions undergoing non-smooth deformations. As mentioned in Sect. 4.3.2.1, shading works on textureless regions and can be used also to infer fine surface details. The *integrated* methods in SfT [27, 48, 51] show that it is possible to combine motion (from textured regions) and shading (from poorly-textured regions) constraints with the physical constraints from the template in order to reconstruct densely at textured and poorly-textured regions. However, since NRSfM does not assume the object's template to be given, combining shading and NRSfM appears to be a much more difficult problem.

4.4 Proposed Solution to NRSfMS

We now focus on the problem of Non-Rigid Structure-from-Motion and Shading (NRSfMS) and we present the first integrated solution. We show in detail how combining motion and shading allows reconstructing creasable, poorly, and well-textured surfaces. The challenge we face is to simultaneously and densely estimate non-smooth, nonrigid shape from each image together with unknown and spatially-varying surface albedo. We solve this with a cascaded initialization and a non-convex refinement that combines a physical, discontinuity-preserving deformation prior with motion, shading, and boundary contour information. Our approach works on both unorganized and organized small-sized image sets, and has been empirically validated on six real-world datasets for which all state-of-the-art approaches fail.

In Sect. 4.4.1, we present our modeling of the problem and our motion and shading-based cost function. In Sect. 4.4.2, we present our optimization framework and in Sect. 4.4.3, we study the basin of convergence of our method and validate it with high-accuracy ground-truth datasets. In Sect. 4.4.4, we provide our conclusions and some research axes of future work.

4.4.1 Problem Modeling

4.4.1.1 Overview

There are many possible ways to define an NRSfMS problem, and many potential choices that must be made regarding scene assumptions, models, unknown, and known terms, *etc.* We present a rigorous definition of NRSfMS through eight fundamental components. To define an NRSfMS problem, we must define or *instantiate* each component. In the following, we describe each component and an instantiation justified by practical considerations for real-world application.

(a) Models (shape, reflectance, illumination, camera response, and camera projection). We use a high-resolution thin-shell 3D mesh to model the object's 3D shape, and a barycentric interpolation to describe deformations across the surface.

This allows us to model complex deformations using high-resolution meshes. Deformation is modeled quasi-isometrically and creases are modeled with a novel implicit energy term as described in Sect. 4.4.1.4. Surface reflectance is modeled using the Lambertian model with piecewise-constant albedo, which has been also used by [5]. This gives a good approximation of many man-made objects such as clothes, fabrics, and cardboards. Scene illumination is assumed to be constant over time and fixed in camera coordinates. In practice, this can be assumed if we have a camera-light rig setup such as an endoscope or camera with flash, or a non-rig where the light and a camera are not physically connected but do not move relative to each other during image acquisition. We use the first-order spherical harmonic model, however, the second-order model can be also used in our proposed solution to increase modeling accuracy. Spherical harmonics are very commonly used in SfS [5, 66] and photometric stereo [34]. We assume the perspective camera model [36], which handles well most real-world cameras, and we assume the camera intrinsics are known a priori using a standard calibration process with, e.g., OpenCV. The camera response model maps the *irradiance image*, i.e., the image which stores the light striking the image plane at each pixel, to the *intensity image*, which is the grayscale image outputted by the camera, before that the camera nonlinearities, such as gamma mapping, vignetting, and digitization, are introduced. For the camera response, we assume that it is linear, which is a valid assumption for many CCD cameras. We assume that it can change over time in order to handle changes due to camera shutter speed and/or exposure. *(b) Known and unknown model parameters.* Most of the above mentioned models have parameters that must be set. The NRSfMS problem changes dramatically according to which model parameters are known *a priori*. We consider the unknowns are as follows: The surface albedo, which, due to the assumption of piecewise-constant albedo, corresponds to an albedo-map segmentation and segment values. The mesh vertex positions in camera coordinates, which provide the 3D shape of the surface in each image. The illumination, the camera responses, and the camera intrinsics are assumed known. These assumptions are reasonable for two reasons. First, the illumination and the camera can be calibrated a priori using standard techniques and the camera responses can be obtained from the camera or computed using, e.g., the background. Second, it is unrealistic to know a priori the reflectance model of a surface as the object is a priori unknown, contrary to SfT. Secondly, as this is the first solution to NRSfMS, our goal is to show that it can be solved in simplified conditions, and in the future, we can investigate releasing the assumption of known illumination, camera response, and intrinsics. *(c) Visual cues.* The visual cues determine which visual information is used to constrain the problem. We use motion, boundary contour, and shading. Motion is used to constrain textured regions of the surface and boundary contours to constrain the perimeter of the surface. We use shading constraint to densely reconstruct surfaces and reveal creases in poorly-textured regions. *(d) Number of required images.* We require at least 5 images. We discuss the implications of using smaller numbers of images in the conclusion Sect. 4.4.3.5. *(e) Expected types of deformations.* We assume quasi-isometric and piecewise-smooth deformations and no tearing. Tearing implies the surface mesh topology must adapt during reconstruction and this adds considerable complexity to the problem. Here it

is sufficient to show that non-torn surfaces can be reconstructed. *(f) Scene geometry.*
We assume the surface to be reconstructed has no self or external occlusions, but
there can be background clutter. These are typical assumptions in NRSfM state-of-
the-art, and the assumption of no occlusions is used to simplify data association (i.e.,
knowing which regions of the images correspond to which regions of the surface).
We also assume there is a *reference image* within the image set. The reference image
is one of the input images that we use to construct the surface's mesh model. We can
use any image as the reference image, however, in practice, we obtain better reconstructions using a reference image where the surface is smooth. *(g) Requirement
for putative correspondences.* Putative correspondences are points in the reference
image whose positions are known in the other images. We assume to know a priori a
set of putative 2D correspondences computed using standard methods such as SIFT.
We assume there may be a small proportion of mismatches, e.g., <20%, which is the
case in real applications. *(h) Surface texture characteristics.* We assume the surface
presents a combination of both well and poorly-textured regions.

4.4.1.2 Shape, Deformation, Reflectance, Illumination, and Camera Modeling

We define Ω as the segmented region of the object of interest in the reference image.
We build the *shape model* by meshing Ω using a regular 2D triangular mesh, with
M vertices and M on the order of 10^4. We denote the mesh's edges as E, where N_E
is the number of edges. Our task is to determine, for each mesh vertex i, its position
$\mathbf{v}_t^i \in \mathbb{R}^3$ in 3D camera coordinates for each image $t \in [1, N]$. We use $\mathcal{V}_t = \{\mathbf{v}_t^i\}_{i \in [1, M]}$
to denote the vertices in 3D camera coordinates for image t. Without loss of generality
we assume the reference image is the first image. We then parameterize $\mathcal{V}_1$ along
lines-of-sight. Specifically, let $\mathbf{u}_i \in \mathbb{R}^2$ denote the 2D position of the ith vertex in
the image, defined in normalized pixel coordinates. Its corresponding position in 3D
camera coordinates at $t = 1$ is $\mathbf{v}_i^1 = d_i[\mathbf{u}_i^\top, 1]^\top$, where d_i is its unknown depth. We
collect these unknown depths into the set $\mathcal{D} = \{d_1, \ldots, d_M\}$. The full set of unknowns
that specify the object's shape in all images is, therefore, $\{\mathcal{D}, \mathcal{V}_2, \ldots, \mathcal{V}_N\}$, which
corresponds to $3M(N - 1) + M$ real-valued unknowns.

The *deformation model* transforms each vertex to 3D camera coordinates: we
model the position of each vertex $i \in \{1, \ldots, M\}$ in camera coordinates by $\mathbf{v}_t^i \in \mathbb{R}^3$,
where t denotes time. We transform a point $\mathbf{u} \in \Omega$ to camera coordinates according
to $\mathcal{V}_t$ with a barycentric interpolation, which is a linear interpolation of the positions
of the three vertices surrounding $\mathbf{u}$. This barycentric interpolation, therefore, defines
a piecewise-linear embedding function from Ω to 3D, parameterized by the vertex
positions $\mathcal{V}_t$. We denote φ this barycentric interpolation and $n(\mathbf{u}; \mathcal{V}_t) : \mathbb{R}^{3 \times M} \to \mathbb{S}_3$
its unit surface normal.

For the *surface reflectance model*, we define an *albedo-map* $A(\mathbf{u}) : \Omega \to \mathbb{R}^+$ as
the function that gives the unknown albedo for a pixel $\mathbf{u} \in \Omega$. From the piecewise-
constant assumption, we can write this as $A(\mathbf{u}) : \Omega \to \mathcal{A}$ where $\mathcal{A} = \{\alpha_1, \ldots, \alpha_K\}$

denotes a discrete set of K unknown albedos with $\alpha_k \in \mathbb{R}^+$. We discuss how A is built in Sect. 4.4.2.

The *illumination model* gives the power and spatial distribution of light. We denote the unknown illumination coefficients by $\mathbf{l}$. This shading equation predicts the intensity of a pixel given the models of illumination, surface shape, surface reflectance, camera projection, and camera response. This starts with the *surface irradiance* which is the amount of light received by the surface. We use the function r to denote the surface irradiance for a normal vector $\mathbf{n}$ according to $\mathbf{l}$. Then, the amount of light reflected by the surface and striking the camera forms the irradiance image. This image contains the photometric variations caused by shading in particular. At any time t, we denote the irradiance image by R_t and the intensity image by L_t. We denote the *camera response function* by $g_t : \mathbb{R} \to \mathbb{R}$ which transforms the irradiance image R_t into the intensity image L_t.

As we use the first-order spherical harmonic model, the illumination model is a combination of a light source at infinity and an ambient term. Note that, as $\mathbf{l}$ is represented by spherical harmonics, the surface irradiance r is linear in $\mathbf{l}$. As we assume the Lambertian reflectance model, we have $r(\mathbf{n}, \mathbf{l}) = (\mathbf{n}^\top, 1) \mathbf{l}$. As we assume g_t is linear, we have $L_t = \beta_t R_t$ with $\beta_t \in \mathbb{R}^+$.

4.4.1.3   Inputs and Outputs

Our inputs are as follows. *(i)* a set of N input RGB images $\{I_t\}_{t\in[1,N]}$, $I_t : \mathbb{R}^2 \to [0, 255]^3$ with a deforming object and the corresponding intensity images $\{L_t\}_{t\in[1,N]}$, $L_t : \mathbb{R}^2 \to \mathbb{R}^+$. In practice, the intensity image L_t is obtained by calibrating radiometrically the camera or by selecting the second component of the projection of the input RGB image I_t in the CIE XYZ color space, which is done for our experiments. *(ii)* the camera intrinsics of all perspective projection functions Π_t. *(iii)* a segmentation of the object of interest in the reference image, denoted by the region $\Omega \subset \mathbb{R}^2$. *(iv)* the scene illumination coefficients $\mathbf{l} \in \mathbb{R}^4$. *(v)* the camera response functions g_t. *(vi)* N sets S_t of matched putative 2D correspondences from Ω to each input image I_t. We denote it by $S_t = \{(\mathbf{u}_j, \mathbf{p}_t^j)\}$ where $\mathbf{u}_j$ denotes the jth 2D point in Ω and $\mathbf{p}_t^j$ denotes its corresponding position in the tth input image I_t. The number of correspondences for each image t is denoted by s_t. Details for how this is done for our experimental datasets are given in Sect. 4.4.3.1.

The outputs of our solution to NRSfMS are: *(i)* the vertices $\mathcal{V}_t$ of the shape model in the camera coordinates for all input images and *(ii)* the segmented albedo-map A with its K segments and values $\{\alpha_1, \ldots, \alpha_K\}$.

4.4.1.4   Problem Modeling with an Integrated Cost Function

The cost function combines *physical deformation priors* (quasi-isometry and smoothing constraints) with shading, motion and boundary constraints extracted from all images. The objective function C_{total} has the following form:

$$C_{total}(\mathcal{V}_1, \ldots, \mathcal{V}_N, \alpha_1, \ldots, \alpha_K) \triangleq \sum_{t=1}^{N} \Big(C_{shade}(\mathcal{V}_t, \alpha_1, \ldots, \alpha_K) + \qquad (4.1)$$

$$\lambda_{motion} C_{motion}(\mathcal{V}_t) + \lambda_{contour} C_{contour}(\mathcal{V}_t) +$$

$$\lambda_{iso} C_{iso}(\mathcal{V}_1, \mathcal{V}_t) + \lambda_{smooth} C_{smooth}(\mathcal{V}_t) \Big).$$

The terms C_{shade}, C_{motion} and $C_{contour}$ are the shading, motion and boundary contour data constraints respectively. The terms C_{smooth} and C_{iso} are the physical deformation prior constraints. The factors λ_{motion}, $\lambda_{contour}$, λ_{iso} and λ_{smooth} are positive weights and are the method's tuning parameters.

The shading constraint. This robustly encodes the Lambertian relationship between surface, albedo, surface irradiance, pixel intensity, and camera response. We use the piecewise-constant albedo model given earlier, and we decide to not optimize all albedo segments. There are two reasons. First, there is a potential difficulty with using shading at textured regions. This comes from the fact that the mis-registration errors at textured regions may imply mis-registration of the albedo-map over the surface. This then may lead to large errors in albedo estimation and surface reconstruction because of the linear dependency of the shading constraint in albedo values. Second, textured regions are very informative for motion constraints. The shading constraint is then less useful or even not useful at textured regions. Therefore, we propose to not use shading in textured regions. For this, we use the fact that textured regions can be detected as small albedo segments. We propose to exclude from the optimization albedo segments which are smaller in area than the threshold T_A (in % of the number of pixels contained in the image). In practice, we found that using $T_A = 0.022\%$ allows to reduce reconstruction errors at textured regions. We give details about how this is integrated to our proposed algorithm in stage 3 in Sect. 4.4.2. We remind that φ is the piecewise-linear embedding function from Ω to 3D, parameterized by the vertex positions, and we form every constraint using this function. We evaluate the shading constraint at each pixel of albedo segments larger than T_A, which gives

$$C_{shade}(\mathcal{V}_t, \alpha_1, \ldots, \alpha_K) \triangleq \frac{1}{|\Omega|} \sum_{\mathbf{u} \in \Omega} \rho_0 \Big(A(\mathbf{u}) \, r \, (n(\mathbf{u}; \mathcal{V}_t); \mathbf{l}) - L_t \, \Pi_t \big(\varphi(\mathbf{u}; \mathcal{V}_t) \big) \Big),$$

$$(4.2)$$

$$\text{with} \quad \rho_0(x) = \begin{cases} \frac{x^2}{2}, & \text{if } |x| \le k \\ k \left(|x| - \frac{k}{2} \right), & \text{if } |x| \ge k, \end{cases} \qquad (4.3)$$

which is the Huber M-estimator. For the experiments, we found that the Huber hyperparameter set to $k = 0.005$ gives the best results. The function ρ is used to enforce similarity between the modeled and measured intensity, while also allowing for some points to violate the model (caused by specular reflection, small shadows, and other unmodeled factors). When the residual of such points is not too high, we find that a robust estimator based on an M-estimator is very effective to handle them. In order to reduce computation time, pixels from Ω are downsampled by a factor of X, by taking one pixel every X pixels from Ω. In practice, we found that using $X = 2$ gives good reconstructions.

The motion constraint. We recall that the set $\mathcal{S}_t$ holds s_t putative correspondences between Ω and image $t \in [1, N]$. The constraint robustly enforces each point $\mathbf{u}_j$ to transform to its corresponding point $\mathbf{p}_t^j$, and is given by

$$C_{motion}(\mathcal{V}_t) \triangleq \sum_{(\mathbf{u}_j, \mathbf{p}_t^j) \in \mathcal{S}_t} \rho_1 \left(\left\| \Pi_t \circ \varphi(\mathbf{u}_j; \mathcal{V}_t) - \mathbf{p}_t^j \right\|_2 \right), \tag{4.4}$$

where ρ_1 is the parameter-free (ℓ_1-ℓ_2) M-estimator

$$\rho_1(x) = 2 \left(\sqrt{1 + \frac{x^2}{2}} - 1 \right). \tag{4.5}$$

This constraint encourages the function φ to project each point $\mathbf{u}_j$ onto the input image at the correspondence position $\mathbf{p}_t^j$.

The boundary contour constraint. We discretize the boundary of Ω to obtain a set of boundary pixels $\mathcal{B} \triangleq \{\mathbf{u}_{k \in [1, N_{\mathcal{B}}]}\}$, with $N_{\mathcal{B}}$ the number of boundary pixels. We then compute a boundariness-map for each image $B_t : \mathbb{R}^2 \to \mathbb{R}^+$ where high values of $B_t(\mathbf{p})$ correspond to a high likelihood of pixel $\mathbf{p}$ being on the boundary contour. The constraint is evaluated as

$$C_{contour}(\mathcal{V}_t) \triangleq \frac{1}{N_{\mathcal{B}}} \sum_{\mathbf{u}_k \in \mathcal{B}} \rho_1 \left(B_t \left(\Pi_t \circ \varphi(\mathbf{u}_k; \mathcal{V}_t); I_t \right) \right). \tag{4.6}$$

From the input image I_t, we build B_t using an edge response filter that is modulated to suppress false positives according to one or more segmentation cues. We use two different segmentation cues: the projection-based and the color-distribution segmentation cues. An illustration of a boundariness-map in given in Fig. 4.4b. The exact choice for computing B_t for each tested dataset in reported in [25].

The quasi-isometry constraint. We enforce quasi-isometry using mesh edge-length constancy. Specifically, we measure the constancy with respect to the mesh edges in the reference image. This is defined as follows:

$$C_{iso}(\mathcal{V}_1, \mathcal{V}_t) \triangleq \frac{1}{|E|} \sum_{(i,j) \in E} \left(1 - \|\mathbf{v}_1^i - \mathbf{v}_1^j\|_2^{-2} \|\mathbf{v}_t^i - \mathbf{v}_t^j\|_2^2 \right)^2. \tag{4.7}$$

This penalizes a change in edge length relative to the mesh in the reference image, and unlike many other ways to impose isometry, is invariant to a global scaling of the reconstruction.

The crease-preserving smoothing constraint. We propose to use from [26], the smoothing constraint based on M-estimators [90]. This will lead to a discontinuity-preserving smoother which automatically deactivates smoothing, where needed at creased regions. Precisely, this constraint penalizes the surface curvature change using a robust bending energy as follows:

$$C_{smooth}(\mathcal{V}_t) \triangleq \frac{1}{|\Omega|} \sum_{\mathbf{u}_j \in \Omega} \rho_1 \left(\frac{\partial^2 \varphi}{\partial \mathbf{u}^2}(\mathbf{u}_j; \mathcal{V}_t) \right). \tag{4.8}$$

In practice, for this constraint, we compute the curvature change in a discrete way. This can be done analytically because position and gradient can be computed using the barycentric coordinates, which is a linear operation in the unknowns, i.e., the vertices. The ability of this constraint to allow creases formation comes from the behavior of the M-estimator for high residuals. Regarding our problem, high residuals in the regularizer correspond to high changes of curvature, which occur at creased regions. Observing the behavior of several M-estimator functions reveal that they grow sub-quadratically at high residuals. Therefore, the impact of high residuals on the optimization of the regularizer will be much smaller when using an M-estimator rather than the ℓ_2 norm, which is used by most of the current methods for the smoothing constraint. It is, however, important to consider that the creases formation is encouraged by the data terms and allowed by the smoothing constraint.

Handling scale. In the cost function (4.1), the shading, the motion, the boundary contour, and the quasi-isometry constraints are invariant to the scale of the reconstruction, however, the smoothing constraint is not invariant. This is because a trivial solution for the smoothing constraint is to put all vertices at the origin. Therefore, to rule out the dependency on scale, we constrain the mean depth of the reconstruction to a fixed positive value. Details are given in Sect. 4.4.3.2.

4.4.2 Optimization Strategy

Optimizing Eq. (4.1) is a nontrivial task because it is large-scale (typically $O(10^5)$ unknowns), is highly non-convex, and the shading constraint requires dense, pixel-level registration. Recall that we do not assume that the images come from an uninterrupted video sequences, which makes dense registration much harder to achieve. We propose a strategy in four stages, illustrated in Fig. 4.3.

Stage 1: We first achieve a rough initial estimate for the shape parameters $(\mathcal{D}, \mathcal{V}_2, \dots, \mathcal{V}_N)$ (and hence an initial estimate for registration) using only motion constraints from the point correspondences. We do this using the initialization-free NRSfM method [14] which has publicly available code.[1] Note that all existing initialization-free surface-based methods assume that the object's surface is smooth in all views, thus the initial estimate will not normally be highly accurate. This provides a rough estimate of the reference image's vertex depths $\mathcal{D}$, which we use to back-project the mesh vertices in the reference image to obtain $\mathcal{V}_1$.

---

[1]The code is available at http://igt.ip.uca.fr/~ab/code_and_datasets/index.php (Matlab SfT Toolbox).

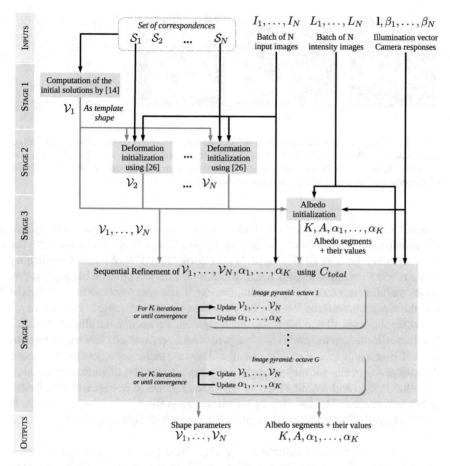

Fig. 4.3 Schematic of our proposed solution to solve NRSfMS

Stage 2: We then use $\mathcal{V}_1$ as a template and perform the SfT method [26] independently on each $\mathcal{V}_t$. It introduces the boundary contour constraints and refines the shape parameters by optimizing Eq. (4.1), with $\lambda_{shade} = 0$, using iterative numerical minimization. [26] also uses two strategies to improve the convergence of the refinement. The first is to refine only with the motion constraint as image data constraint, then we add the boundary contour constraint. The second is to construct from each input image I_t the boundariness-map B_t using an image pyramid, and sequentially optimize with each pyramid level. We found that three octaves for the pyramid level provide good convergence.

Stage 3: This consists of the segmentation of the reference image I_1 in regions of constant albedos and in the estimation of the albedos by inverting the shading equation. For this, we use an intrinsic image decomposition [10], on the reference image's intensity image and cluster the resulting "reflectance image" using [24], with a low

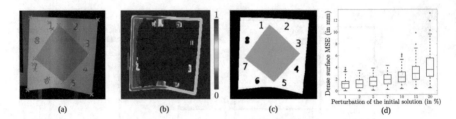

(a)   (b)   (c)   (d)

Fig. 4.4 a Visualization of the correspondences of the input image n° 1 (zoom) of the *paper fortune teller* dataset. **b** Boundariness-map (zoom) for the input image n° 1 of the *paper fortune teller* dataset. **c** Albedo-map (zoom) estimated for the *paper fortune teller* dataset. **d** Numerical results of convergence basin analysis for the *paper fortune teller* dataset

cluster tolerance (we use a default of 10). For each cluster k, we assign a corresponding albedo α_k: for each pixel $\mathbf{u}_j$ in the cluster, we estimate its albedo by inverting the shading equation: $\alpha \approx L_t \left(\Pi_t \circ \varphi(\mathbf{u}; {}'V_t) \right) r \left(n(\mathbf{u}; {}'V_t); \mathbf{l} \right)^{-1}$. We then initialize α_k as the median overall estimates within the cluster. This can be done because, at this stage, we have estimated the scene illumination, the camera response for each image and the shape parameters. We aim for an oversegmentation: neighboring segments can share the same albedo but within each segment we assume the albedo constant. The reason is that our method is not designed to recover from an under segmentation. Even if oversegmentation requires more unknowns, under segmentation is a more difficult problem since it may strongly impact the estimation of surface orientation and illumination and requires then an automatic process to re-segment the albedo-map when needed. The last step of the clustering is the thresholding of the pixels number of each albedo segment to remove the ones which correspond to the textured regions.

Figure 4.4c shows an illustration of a segmented albedo-map: the black holes visible on the surface corresponds to the textured regions whose area is smaller than T_A. The black holes visible on the surface in Fig. 4.4c corresponds to these textured regions. If there are K segments, then the albedo set $\{\alpha_1, \ldots, \alpha_K\}$ has size K.

Stage 4: We refine alternately the shape parameters and the albedo values by minimizing Eq. (4.1) using all constraints. This is achieved with Gauss-Newton iterative optimization and backtracking line-search. Because of the very large number of unknowns, at each iteration, we solve the normal equations using an iterative solver (diagonally-preconditioned conjugate gradient), with a default iteration limit of 200. Recall that there is a scale ambiguity (as in all NRSfM problems), because we cannot differentiate a smaller surface viewed close to the camera from a large surface viewed far away. We fix the scale ambiguity by scaling all vertices to have a mean depth of 1 after each iteration. To achieve good convergence, we use the two strategies of [27]. First, we use only the motion constraint as image data constraint to refine, then we add the boundary contour constraint and end by refining the three image data constraints. Second, we construct from each input image I_t the boundariness-map B_t using an image pyramid, blur each L_t with a Gaussian blur pyramid, and sequentially optimize with each pyramid level, with a default of three octaves. For the two first levels Gaussian blur pyramid, the kernel sizes, and standard-deviations are respectively $h_1 = (10, 10)$ and $\sigma_1 = 5$ and $h_2 = (5, 5)$ and $\sigma_2 = 2.5$. At the finest level, we do not apply any Gaussian blur. For the three pyramid levels, we run Gauss-Newton until either convergence is reached (with the total cost difference between two consecutive iterations being strictly lower than $1e-4$) or a fixed number of iterations have passed (we use $\kappa = 20$ iterations).

4.4.3 Experimental Validation

We divide the experimental validation into two parts. First, we analyze the convergence basin of our energy function through perturbation analysis. This is to understand how sensitive our formulation is to the initial solution, and fundamentally, whether the NRSfMS problem can be cast as an energy-based minimization with a strong local minimum near the true solution. Second, we compare performance to state-of-the-art NRSfM methods, using six datasets, all with ground-truth.

4.4.3.1 Methods Compared and Datasets

We compare with the following competitive NRSfM methods [14, 60, 81], denoted, respectively, with **Va09**, **Ch14**, **Pa16**. We compare to these methods because they reconstruct dense surfaces. To see the contribution of some constraints of Eq. (4.1), we compare with four versions of our method, **NoS**, where shading is not used, **NoB**, where the boundary constraint is not used in stages 2 and 4, **NoI**, where the quasi-isometry constraint is not used in stages 2 and 4, and **NoSm**, where the smoothing constraint is not used in stages 2 and 4.

We evaluated on six real-world datasets which mostly respect the Lambertian assumption: *floral paper* and *paper fortune teller* from [27], *creased paper*, *pillow cover* and *hand bag* from [28] and *Kinect paper* from [80]. Each dataset consists of a disc-topology surface in 5 different deformed states, with one state per image. We show them in Figs. 4.6 and 4.7. The five first datasets have the following conditions: *(i)* the object has a poorly-textured surface, *(ii)* several images show the surface creased, *(iii)* a highly-accurate depth-map associated with each image, *(iv)* the illumination vector is in 3D camera coordinates. These five datasets have been acquired with the structured light system [19]. As the *Kinect paper* has no accompanying illumination parameters and no camera response function, these are computed prior to the reconstruction. More details are given in [25]. Each dataset has a set of point correspondences between the first and all other images. As all datasets, except the *Kinect paper* dataset, are poorly-textured, the correspondences are sparse. We note that manual correspondences are commonly used to evaluate NRSfM methods and this is why the correspondences of our datasets were computed manually. These correspondences are distinctive points such as the texture discontinuities along the printed numbers of the *paper fortune teller* dataset, visible in Fig. 4.4a. The datasets *floral paper, paper fortune teller, creased paper, pillow cover.* and *hand bag* have, respectively, 20, 24, 20, 69, and 155 correspondences and their image size is 1288×964 pixels. The *Kinect paper* dataset presents images with 640×480 pixels and 1503 correspondences computed by [29]. The datasets *floral paper, paper fortune teller, creased paper, pillow cover,* and *hand bag* are publicly available.[2]

4.4.3.2 Implementation Details and Evaluation Metrics

We constructed, for all experiments, the reference meshes by laying a triangulated 100×100 vertex regular grid on the reference image which was then cropped to Ω. We also discretized the boundary points of the texture-map to $N_{\mathcal{B}} = 1000$ uniformly spaced points. For the compared methods, there is no way to automatically optimize their hyperparameters. We then tried our best to do this by hand, to obtain the best reconstruction accuracy on all datasets. For our method, all experiments were ran using the same hyperparameters, which were manually set. In Appendix Sect. 4.5, Tables 4.1 and 4.2 give the weights of the different constraints and the hyperparameters for our method and the compared methods.

To measure reconstruction accuracy, we compared 3D distances and normals with respect to ground-truth using, respectively, the Mean Shape Error (MSE) and the Mean Normal Error (MNE). To investigate the contribution of the shading constraint, this was done at two locations: *(i)* densely across the ground-truth surface, and *(ii)* densely at creased regions, which are any points on the ground-truth surfaces that are within 5 mm of a surface crease. Both grids were constructed by sampling uniformly the respective locations. Because reconstruction is up to scale, we computed for each method the best-fitting scale factor that aligns the predicted point correspondences with their true locations in the ℓ_2 sense, then measured accuracy with the scale-corrected reconstruction.

4.4.3.3 Convergence Basin Analysis

We performed this with perturbation analysis as follows. We started with an initial reconstruction close to the ground-truth, then applied a low-pass filter (to smooth out creases, as we do not expect them in the initial solution), and randomly perturbed the vertex positions using smooth deformation functions. For each perturbation, we optimized Eq. (4.1), by performing stages 3 and 4 in Sect. 4.4.2. The perturbation was implemented using a $4 \times 4 \times 4$ B-spline enclosing the reconstructed surfaces and randomly perturbing the spline control points at 7 different noise levels, with 30 random perturbations per noise level. Figure 4.4d reports results as box-plots for the *paper fortune teller* dataset. The x-axis gives the average perturbation in % for each noise level from the initial solution. The y-axis gives the dense surface MSE for each random sample. Similar results are obtained for the *floral paper* and *creased paper*

---

[2]The datasets are available at http://igt.ip.uca.fr/~ab/code_and_datasets/index.php.

datasets and are reported in [25]. For small noise levels ($<5\%$), the box-plots are very similar, which tells us our energy landscape has a strong local minimum close to the ground-truth. This supports our claim that the NRSfMS problem can be cast as an energy-based minimization (via Eq. (4.1)). For larger noise levels ($>5\%$), we can see a significant increase in error, indicating that the optimization now becomes trapped more frequently in local minima.

4.4.3.4 Quantitative and Qualitative Results

We show in Figs. 4.6 and 4.7, the six datasets and their reconstructions from our method and the best performing previous method (the one with lowest MSE with respect to *(ii)* above). Visually, we can see that considerable surface detail is accurately reconstructed by our method, as well as the global shape. In Fig. 4.5, we give the reconstruction accuracy statistics across all test datasets and all compared methods. The *Kinect paper* dataset has no creases and the deformation is very smooth in all images. We observe that, for all datasets other than *Kinect paper*, there is a good improvement with respect to all error metrics compared to the other methods. For the *Kinect paper* dataset, we see that our method does not obtain the highest accuracy across all error metrics. The reason is that it is a very smooth, densely-textured surface, and shading is not needed to achieve an accurate reconstruction. However, our method still obtains competitive results on this dataset. We observe that the use of shading improves globally the shape of the reconstructions and that the boundary contour constraint allows using shading better. The reduced performance with **NoSm** confirms that the smoothing constraint acts as a regularizer. An observation of the 3D surfaces reconstructed without the smoothing constraint shows that the creases cannot be formed and the surfaces are very smooth. Figure 4.5 does not show the results for **NoI** because, for every dataset, the surfaces collapse to the origin during stage 2. This is consistent with the fact that isometry constraint makes the problem well-posed as it has been mentioned in Sect. 4.3.1.1. In Appendix Sect. 4.5, Table 4.3 gives the processing times for our method and the three compared methods with respect to each dataset.

4.4.3.5 Limitations and Failure Modes

We discuss here the main limitations and the failure modes of our solution to NRSfMS. Our method is limited by the assumptions made in Sect. 4.4.1. These are that we have isometric deformations, piecewise-constant albedo, fixed and known illumination vector, and known camera responses. One important limitation of our solution is the parameter tuning since the parameters of our method are set manually and may vary with the datasets. This is because we observe that we did not find default parameters for all datasets yielding to the best reconstruction accuracy. It would be interesting to investigate whether there exist fixed tuning parameters which work well on all datasets, using, e.g., grid search. Our approach regarding

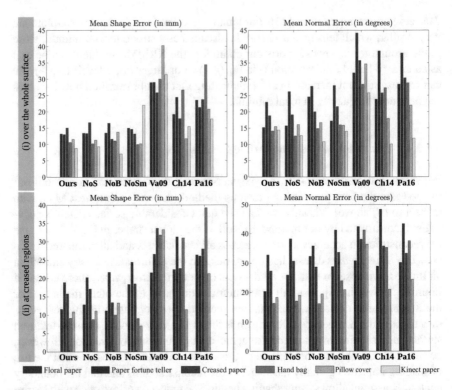

Fig. 4.5 Reconstruction accuracy statistics across all test datasets and all compared methods. Also, the *Kinect paper* dataset does not present any crease

the estimation of the albedo segments and their values present a drawback. Our method uses a single image, the reference image, to estimate the albedo segments and it cannot split or merge them during the following steps because of our modeling given in Sect. 4.4.1.2. Particularly, this may be problematic when some deformations occurring in the reference image lead the albedo initialization to merge two albedo segments with significantly different values. The constraint on the fixed number of albedo segments can be relaxed and mechanisms to automatically adjust them during the refinement step can be studied, such as the cost term of [33] which encourages piecewise-constant albedo segments by penalizing gradients on the albedo value through a ℓ_0 norm. Another limitation is that we perform our experiments with batches of 5 images and we have not performed a theoretical analysis to establish the minimum number of images to solve NRSfMS. Some failure modes are caused by the joint use of shading and motion visual cues. The first one is that, as in SfS, our method may then suffer from localized convex/concave ambiguities, which tends to worsen flawed initial solutions. The second failure mode is the under segmentation of the albedo-map, which may lead to incorrect surface orientation. The third one is the presence of some false positive creases, as Figs. 4.6 and 4.7, show. This failure mode is linked to the first one, but is more general and can integrate other sources

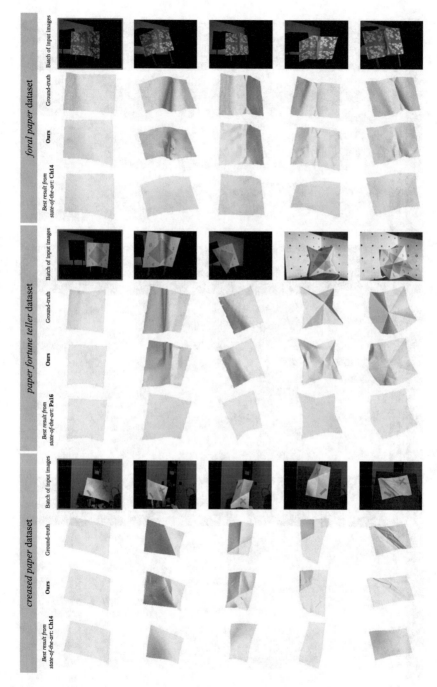

Fig. 4.6 Renderings for the *floral paper*, the *paper fortune teller* and *creased paper* datasets with ground-truth. Here we show the images from each dataset, and sample reconstructions from one of the images using our method and the best performing NRSfM method. We frame the reference image in blue

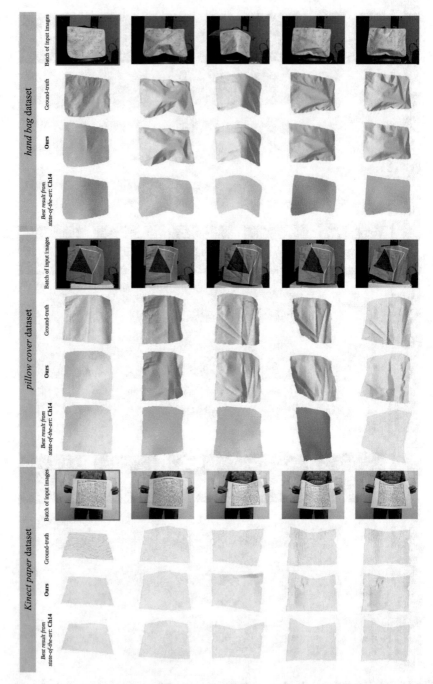

Fig. 4.7 Renderings for the *hand bag*, the *pillow cover* and *Kinect paper* datasets with ground-truth. Here we show the images from each dataset, and sample reconstructions from one of the images using our method and the best performing NRSfM method. We frame the reference image in blue

of errors such as misregistration or the robust estimator applied in the shading constraint. The last failure mode, which is not caused by the use of shading, is when the
initial solutions given by stage 1 are not reliable. Typically this occurs if there are
very few, poorly-distributed point correspondences. In these cases, it is difficult to
initialize dense shape with any current SfT method. For unorganized image sets, this
is a difficult problem to overcome. For video sequences, dense point correspondences
can usually be obtained by exploiting temporal continuity and dense frame-to-frame
tracking [17].

4.4.4 Conclusion and Open Problems

We have presented the first study of the NRSfMS problem as an illustration of the
combination of motion and shading visual cues to infer the 3D shape of deforming
objects from a single camera. NRSfMS does not assume the 3D geometry of the
surface is known prior to reconstruction, and solves the problem using only a set
of images and models pertaining to camera projection, scene illumination, surface
reflectance, and camera response. We have shown for the first time that it is possible
to solve NRSfMS when some of the model parameters are unknown (specifically
surface reflectance) and solve jointly with reconstruction. NRSfMS is a hard and
important vision problem, needed for high-accuracy dense reconstruction of poorly-
textured surfaces undergoing non-smooth deformation from 2D images. We have
proposed an energy-based solution and a cascaded numerical optimization strategy,
and have shown encouraging results on six real-world datasets, for which all competitive NRSfM methods fail. This marks the first time that strongly creased, deformable,
poorly-textured surfaces with unknown albedos have been densely reconstructed and
registered from 2D image sets without shape prior knowledge on the object.

There are many possible future directions for NRSfMS. These involve both theoretical analysis to understand the problem well-posedness, and explorations to release
some of the practical limits of our approach as given in Sect. 4.4.3.5. Regarding the
former, NRSfM can be solved up to ambiguities with two images and SfS can be
solved with one image when the illumination and the surface reflectance are known.
At first sight, two images seem to be sufficient to solve NRSfMS, however, a thorough
theoretical study would be required for the version of NRSfMS presented in Sect. 4.4
and also for the other possible instantiations of NRSfMS.

Regarding the latter, we propose two main research directions. The first direction
is to consider more uncontrolled settings and examine other strategies to use motion
in order to improve the use of shading. Examples of these settings are when the scene
illumination is not known a priori or when the relative position between the camera
and the light source changes over time. This will require a careful theoretical study
of well-posedness, innovative initialization and optimization strategies. Our work
shows that motion can provide an accurate reconstruction of surface normals at well-
textured regions, and these regions can be used to estimate photometric parameters. It
may be possible to estimate other photometric parameters such as camera response,

using data at the reconstructed textured regions. A second direction considers occlusions and shadows. Some solutions have been proposed for handling occlusions in SfT [17, 52, 57] and external occlusions in NRSfM [60], and for handling shadows in SfS [22]. However, there is no attempt to reason simultaneously about occlusions and shadows in NRSfMS, which is required to achieve robust reconstruction in the wild. Reasoning first with the correspondence constraints, i.e., features motion over the surface may provide more robust initial solutions which can be then more easily refined by reasoning with shading constraints.

4.5 Appendix

Tables 4.1 and 4.2 give the hyperparameters which we used to produce the results on the six datasets used in Sect. 4.4.3. We denote the SfT method [26], used in stage 2 of our NRSfMS method, with **Ga16**.

Table 4.3 gives the average processing times for our NRSfMS method and the compared methods for each dataset. We refer to Sect. 4.4.3.2 for the implementation details and Sect. 4.4.3.1 for the dataset details.

Table 4.1 Hyperparameter values used to evaluate our NRSfMS method

		floral paper	paper fortune teller	creased paper	pillow cover	hand bag	kinect paper
Ga16	M	1e4					
	$N_{\mathcal{B}}$	1e3					
	$\lambda_{contour}$	1e−5	4e−4	4e−4	4e−4	4e−4	0.04
	λ_{iso}	4e−4	0.16	4e−3	4e−3	0.04	0.04
	λ_{smooth}	6e−15	2.4e−13	1.6e−14	1.6e−14	1.6e−14	4e−13
Ours	M	1e4					
	$N_{\mathcal{B}}$	1e3					
	k_{shade}	5e−3	5e−3	5e−3	5e−3	5e−3	5e−3
	λ_{motion}	0.088	0.154	1	10	10	10
	$\lambda_{contour}$	1.25e−4	0.011	0.01	1.67e−4	1.67e−4	1.67e−4
	λ_{iso}	3.8e−3	0.025	0.167	0.167	0.167	0.5
	λ_{smooth}	2.5e−12	9.2e−12	3.33e−11	3.33e−11	3.33e−11	8.33e−10

Table 4.2 Hyperparameter values used to evaluate all compared NRSfM methods

		floral paper	paper fortune teller	creased paper	pillow cover	hand bag	kinect paper
Va09	depth.nC	30	30	30	28	30	30
	depth.er	6	0.06	0.2	8	0.2	6
	embedding.nC	30					
	embedding.er	0.01	1e−6	1e−6	1e−4	1e−6	0.01
	homographies.neigh	100					
Ch14	depth.nC	28	30	28	16	28	30
	depth.er	5	1	0.7	1	8	0.9
	warps.nC	28	20	28	16	28	30
	warps.er	0.01	1e−3	9e−4	1e−4	1e−3	0.01
	homographies.neigh	40	40	40	80	40	40
Pa16	schwarzianParam	2e−5					1e−3
	warps.nC	60					
	warps.er	1e−4					
	depth.nC	100					
	depth.er	1					10

Table 4.3 Processing time in minutes for our method and the three compared methods with respect to each dataset. For the compared methods, we explain the differences in time for the *Kinect paper* dataset because of the number of correspondences which is significantly larger than the other five datasets

	floral paper	paper fortune teller	creased paper	pillow cover	hand bag	kinect paper
Ours	28'39	41'40	20'4	45'23	34'47	35'58
Va09	0'6	0'5	0'6	0'6	0'58	11'11
Ch14	0'18	0'15	0'14	0'8	0'29	1'33
Pa16	9'41	14'10	9'19	7'15	12'38	30'54

References

1. 3Dflow (2017) 3DF Zephyr. https://www.3dflow.net
2. Agisoft (2014) PhotoScan version 1.2.3 build 2331. http://www.agisoft.com
3. Ahmed AH, Farag AA (2007) Shape from shading under various imaging conditions. In: CVPR
4. Akhter I, Sheikh Y, Khan S, Kanade T (2009) Nonrigid structure from motion in trajectory space. In: NIPS
5. Barron JT, Malik J (2015) Shape, illumination, and reflectance from shading. IEEE Trans Pattern Anal Mach Intell 37(8):1670–1687
6. Bartoli A, Gérard Y, Chadebecq F, Collins T, Pizarro D (2015) Shape-from-template. IEEE Trans Pattern Anal Mach Intell 37(10):2099–2118
7. Bartoli A, Özgür E (2016) A perspective on non-isometric shape-from-template. In: ISMAR
8. Beeler T, Bickel B, Beardsley P, Sumner B, Gross M (2010) High-quality single-shot capture of facial geometry. In: SIGGRAPH

9. Belhumeur PN, Kriegman DJ, Yuille AL (1997) The bas-relief ambiguity. In: CVPR
10. Bell S, Bala K, Snavely N (2014) Intrinsic images in the wild. ACM Trans Graph (SIGGRAPH) 33(4)
11. Bregler C, Hertzmann A, Biermann H (2000) Recovering non-rigid 3D shape from image streams. In: CVPR
12. Brostow GJ, Hernández C, Vogiatzis G, Stenger B, Cipolla R (2001) Video normals from colored lights. IEEE Trans Pattern Anal Mach Intell 33(10):2104–2114
13. Brunet F, Hartley R, Bartoli A (2014) Monocular template-based 3D surface reconstruction: convex inextensible and nonconvex isometric methods. Comput Vis Image Underst 125:138–154 August
14. Chhatkuli A, Pizarro D, Bartoli A (2014) Non-rigid shape-from-motion for isometric surfaces using infinitesimal planarity. In: BMVC
15. Chhatkuli A, Pizarro D, Bartoli A, Collins T (2017) A stable analytical framework for isometric shape-from-template by surface integration. IEEE Trans Pattern Anal Mach Intell 39(5):833–850
16. Chhatkuli A, Pizarro D, Collins T, Bartoli A (2016) Inextensible non-rigid shape-from-motion by second-order cone programming. In: CVPR
17. Collins T, Bartoli A (2015) Realtime shape-from-template: system and applications. In: ISMAR
18. Dai Y, Li H, He M (2014) A simple prior-free method for non-rigid structure-from-motion factorization. Int J Comput Vis 107(2):101–122
19. David 3D Scanner (2014) http://www.david-3d.com/en/products/david4
20. Durou J-D, Falcone M, Sagona M (2008) Numerical methods for shape-from-shading: a new survey with benchmarks. Comput Vis Image Underst 109(1):22–43 January
21. Ecker A, Jepson AD (2010) Polynomial shape from shading. In: CVPR
22. Falcone M, Sagona M, Seghini A (2003) A scheme for the shape-from-shading model with "black shadows". Numerical mathematics and advanced applications, pp 503–512
23. Fayad J, Agapito L, Del Bue A (2010) Piecewise quadratic reconstruction of non-rigid surfaces from monocular sequences. In: ECCV
24. Fukunaga K, Hostetler L (1975) The estimation of the gradient of a density function, with applications in pattern recognition. IEEE Trans Inf Theory 21(1):32–40
25. Gallardo M (2018) Contributions to monocular deformable 3D reconstruction: curvilinear objects and multiple visual cues. Theses, Université Clermont Auvergne. https://tel.archives-ouvertes.fr/tel-01930477
26. Gallardo M, Collins T, Bartoli A (2016) Can we jointly register and reconstruct creased surfaces by shape-from-template accurately? In: ECCV
27. Gallardo M, Collins T, Bartoli A (2016) Using shading and a 3D template to reconstruct complex surface deformations. In: BMVC
28. Gallardo M, Collins T, Bartoli A (2017) Dense non-rigid structure-from-motion and shading with unknown albedos. In: ICCV
29. Garg R, Roussos A, Agapito L (2013) Dense variational reconstruction of non-rigid surfaces from monocular video. In: CVPR
30. Gotardo PFU, Martinez AM (2011) Kernel non-rigid structure from motion. In: ICCV
31. Gotardo PFU, Simon T, Sheikh Y, Matthews I (2015) Photogeometric scene flow for high-detail dynamic 3D reconstruction. In: ICCV
32. Gumerov N, Zandifar A, Duraiswami R, Davis LS (2004) Structure of applicable surfaces from single views. In: ECCV
33. Haefner B, Quéau Y, Möllenhoff T, Cremers D (2018) Fight ill-posedness with ill-posedness: single-shot variational depth super-resolution from shading. In: CVPR
34. Haefner B, Ye Z, Gao M, Wu T, Quéau Y, Cremers D (2019) Variational uncalibrated photometric stereo under general lighting. In: ICCV
35. Haouchine N, Dequidt J, Berger MO, Cotin S (2014) Single view augmentation of 3D elastic objects. In: ISMAR
36. Hartley RI, Zisserman A (2003) Multiple view geometry in computer vision, 2nd edn. Cambridge University Press, Cambridge

37. Horn BKP (1970) Shape from shading: a method for obtaining the shape of a smooth shape of a smooth opaque object from one view. Technical report, Cambridge, MA, USA
38. Horn BKP (1989) Shape from shading, pp 123–171
39. Ikeuchi K, Horn BKP (1981) Numerical shape from shading and occluding boundaries. Artif Intell 17(1):141–184
40. Innmann M, Zollhöfer M, Nießner M, Theobalt C, Stamminger M (2016) VolumeDeform: real-time volumetric non-rigid reconstruction. In: ECCV
41. Jin H, Cremers D, Wang D, Prados E, Yezzi A, Soatto S (2008) 3-D reconstruction of shaded objects from multiple images under unknown illumination. Int J Comput Vis 76(3):245–256
42. Kim K, Torii A, Okutomi M (2016) Multi-view inverse rendering under arbitrary illumination and albedo. In: ECCV
43. Kimmel R, Bruckstein AM (1994) Global shape-from-shading. In: ICPR
44. Koo B, Özgür E, Le Roy B, Buc E, Bartoli A (2017) Deformable registration of a preoperative 3D liver volume to a laparoscopy image using contour and shading cues. In: MICCAI
45. Langguth F, Sunkavalli K, Hadap S, Goesele M (2016) Shading-aware Multi-view Stereo. In: ECCV
46. Lee KM, Kuo C-C (1997) Shape from shading with a generalized reflectance map model. Comput Vis Image Underst 67(2):143–160
47. Lee KM, Kuo CCJ (1993) Shape from shading with a linear triangular element surface model. IEEE Trans Pattern Anal Mach Intell 15(8):815–822
48. Liu-Yin Q, Yu R, Agapito L, Fitzgibbon A, Russell C (2016) Better together: joint reasoning for non-rigid 3D reconstruction with specularities and shading. In: BMVC
49. Lowe DG (2004) Distinctive image features from scale-invariant keypoints. Int J Comput Vis 60(2):91–110
50. Magnenat S, Ngo DT, Zund F, Ryffel M, Noris G, Rothlin G, Marra A, Nitti M, Fua P, Gross M, Sumner R (2015) Live texturing of augmented reality characters from colored drawings. IEEE Trans Vis Comput Graph 21(11):1201–1210
51. Malti A, Bartoli A (2014) Combining conformal deformation and cook-torrance shading for 3D reconstruction in laparoscopy. IEEE Trans Biol Eng 61(6):1684–1692
52. Malti A, Bartoli A, Collins T (2011) A pixel-based approach to template-based monocular 3D reconstruction of deformable surfaces. In: Proceedings of the IEEE international workshop on dynamic shape capture and analysis at ICCV
53. Malti A, Bartoli A, Hartley RI (2015) A linear least-squares solution to elastic shape-from-template. In: CVPR
54. Moreno-Noguer F, Salzmann M, Lepetit V, Fua P (2009) Capturing 3D stretchable surfaces from single images in closed form. In: CVPR
55. Newcombe RA, Fox D, Seitz SM (2015) DynamicFusion: reconstruction and tracking of non-rigid scenes in real-time. In: CVPR
56. Ngo TD, Östlund J, Fua P (2016) Template-based monocular 3D shape recovery using laplacian meshes. IEEE Trans Pattern Anal Mach Intell 38(1):172–187
57. Ngo TD, Park S, Jorstad AA, Crivellaro A, Yoo C, Fua P (2015) Dense image registration and deformable surface reconstruction in presence of occlusions and minimal texture. In: ICCV
58. Okatani T, Deguchi K (1996) Shape reconstruction from an endoscope image by SfS technique for a point light source at the projection Center. Comput Vis Image Underst 66(2):119–131 July
59. Özgür E, Bartoli A (2017) Particle-SfT: a provably-convergent, fast shape-from-template algorithm. Int J Comput Vis 123(2):184–205 June
60. Parashar S, Pizarro D, Bartoli A (2016) Isometric non-rigid shape-from-motion in linear time. In: CVPR
61. Pentland AP (1984) Local shading analysis. IEEE Trans Pattern Anal Mach Intell 6(2):170–187
62. Pentland AP (1988) Shape information from shading: a theory about human perception. In: ICCV
63. Perriollat M, Hartley R, Bartoli A (2011) Monocular template-based reconstruction of inextensible surfaces. Int J Comput Vis 95(2):124–137 November

64. Prados E, Faugeras O (2005) Shape from shading: a well-posed problem? In: CVPR
65. Pumarola A, Agudo A, Porzi L, Sanfeliu A, Lepetit V, Moreno-Noguer F (2018) Geometry-aware network for non-rigid shape prediction from a single view. In: CVPR
66. Quéau Y, Mélou J, Castan F, Cremers D, Durou J (2017) A variational approach to shape-from-shading under natural illumination. In: EMMCVPR
67. Richardson E, Sela M, Kimmel R (2016) 3D face reconstruction by learning from synthetic data. In: 3DV
68. Richter SR, Roth S (2015) Discriminative shape from shading in uncalibrated illumination. In: CVPR
69. Rouy E, Tourin A (1992) A viscosity solutions approach to shape-from-shading. SIAM J Numer Anal 29(3):867–884 June
70. Salzmann M, Fua P (2009) Reconstructing sharply folding surfaces: a convex formulation. In: CVPR
71. Salzmann M, Fua P (2011) Linear local models for monocular reconstruction of deformable surfaces. IEEE Trans Pattern Anal Mach Intell 33(5):931–944
72. Salzmann M, Pilet J, Ilic S, Fua P (2007) Surface deformation models for nonrigid 3D shape recovery. IEEE Trans Pattern Anal Mach Intell 29(8):1481–1487
73. Sorkine O, Alexa M (2007) As-rigid-as-possible surface modeling. In: Symposium on geometry processing, pp 109–116
74. Sundaram N, Brox T, Keutzer K (2010) Dense point trajectories by GPU-accelerate large displacement optical flow. In: ECCV
75. Tankus A, Sochen N, Yeshurun Y (2005) Shape-from-shading under perspective projection. Int J Comput Vis 63(1):21–43 June
76. Taylor J, Jepson AD, Kutulakos K (2010) Non-rigid structure from locally-rigid motion. In: CVPR
77. Thies J, Zollhöfer M, Stamminger M, Theobalt C, Nießner M (2016) Face2Face: real-time face capture and reenactment of RGB videos. In: CVPR
78. Tsai P-S, Shah M (1994) Shape from shading using linear approximation. Image Vis Comput 12(8):487–498
79. Valgaerts L, Wu C, Bruhn A, Seidel H-P, Theobalt C (2012) Lightweight binocular facial performance capture under uncontrolled lighting. In: SIGGRAPH
80. Varol A, Salzmann M, Fua P, Urtasun R (2012) A constrained latent variable model. In: CVPR
81. Varol A, Salzmann M, Tola E, Fua P (2009) Template-free monocular reconstruction of deformable surfaces. In: ICCV
82. Varol A, Shaji A, Salzmann M, Fua P (2012) Monocular 3D reconstruction of locally textured surfaces. IEEE Trans Pattern Anal Mach Intell 34(6)
83. Vicente S, Agapito L (2012) Soft inextensibility constraints for template-free non-rigid reconstruction. In: ECCV
84. Vicente S, Agapito L (2013) Balloon shapes: reconstructing and deforming objects with volume from images. In: 3DV
85. Wang X, Salzmann M, Wang F, Zhao J (2016) Template-free 3D reconstruction of poorly-textured nonrigid surfaces. In: ECCV
86. Wu C, Narasimhan SG, Jaramaz B (2010) A multi-image shape-from-shading framework for near-lighting perspective endoscopes. Int J Comput Vis 86(2):211–228
87. Wu C, Varanasi K, Liu Y, Seidel HP, Theobalt C (2011) Shading-based dynamic shape refinement from multi-view video under general illumination. In: ICCV
88. Xiong Y, Chakrabarti A, Basri R, Gortler SJ, Jacobs DW, Zickler T (2015) From shading to local shape. IEEE Trans Pattern Anal Mach Intell 37(1)
89. Yu R, Russell C, Campbell NDF, Agapito L (2015) Direct, dense, and deformable: template-based non-rigid 3D reconstruction from RGB video. In: ICCV
90. Zhang Z (1997) Parameter estimation techniques: a tutorial with application to conic fitting. Image Vis Comput 15(1):59–76
91. Zhou Z, Wu Z, Tan P (2013) Multi-view photometric stereo with spatially varying isotropic materials. In: CVPR

Chapter 5
On the Well-Posedness of Uncalibrated Photometric Stereo Under General Lighting

Mohammed Brahimi, Yvain Quéau, Bjoern Haefner, and Daniel Cremers

Abstract Uncalibrated photometric stereo aims at estimating the 3D-shape of a surface, given a set of images captured from the same viewing angle, but under unknown, varying illumination. While the theoretical foundations of this inverse problem under directional lighting are well-established, there is a lack of mathematical evidence for the uniqueness of a solution under general lighting. On the other hand, stable and accurate heuristical solutions of uncalibrated photometric stereo under such general lighting have recently been proposed. The quality of the results demonstrated therein tends to indicate that the problem may actually be well-posed, but this still has to be established. The present paper addresses this theoretical issue, considering first-order spherical harmonics approximation of general lighting. Two important theoretical results are established. First, the orthographic integrability constraint ensures uniqueness of a solution up to a global concave–convex ambiguity, which had already been conjectured, yet not proven. Second, the perspective integrability constraint makes the problem well-posed, which generalizes a previous result limited to directional lighting. Eventually, a closed-form expression for the unique least-squares solution of the problem under perspective projection is provided, allowing numerical simulations on synthetic data to empirically validate our findings.

M. Brahimi (✉) · B. Haefner · D. Cremers
Technical University of Munich, Munich, Germany
e-mail: mohammed.brahimi@tum.de

B. Haefner
e-mail: bjoern.haefner@tum.de

D. Cremers
e-mail: cremers@tum.de

Y. Quéau
Normandie Université, UNICAEN, ENSICAEN, CNRS, Caen, France
e-mail: yvain.queau@ensicaen.fr

© Springer Nature Switzerland AG 2020   147
J.-D. Durou et al. (eds.), *Advances in Photometric 3D-Reconstruction*,
Advances in Computer Vision and Pattern Recognition,
https://doi.org/10.1007/978-3-030-51866-0_5

5.1   Introduction

Among the many photographic techniques which can be considered for the 3D-reconstruction of a still surface, photometric stereo [43] is often considered as a first choice when it comes to the recovery of very thin geometric structures. Nevertheless, the classic formulation of photometric stereo requires illumination to be highly controlled: each image must be captured under a single collimated light source at infinity, and the direction and relative intensity of each source must be calibrated beforehand. In practice, this restricts possible applications of the technique to laboratory setups where collimation of light can be ensured and a (possibly tedious) calibration procedure can be carried out.

Considering uncalibrated general lighting, i.e., lighting induced by unknown, non-collimated sources and in the presence of ambient lighting, would both drastically simplify the 3D-scanning process for nonexperts, and allow to bring photometric stereo outside of the lab [37]. The theoretical foundations of the problem under uncalibrated directional lighting are well-understood: the solution can be recovered only up to a linear transformation [12]. When integrability is enforced, this linear ambiguity reduces to the generalized bas-relief one under orthographic projection [44], and vanishes under perspective projection [25]. This work rather focuses on uncalibrated general lighting represented using first-order spherical harmonics [2, 35], in which case the solution can be recovered only up to a Lorentz transformation [3] and it has been conjectured—but not proven yet, that additional constraints such as integrability may reduce this ambiguity. One reason for thinking that this conjecture might hold is that stable numerical implementations of uncalibrated photometric stereo under general illumination have been proposed recently, under both orthographic [1, 23] and perspective [11] projections. Despite having no theoretical foundation, the results provided therein do not exhibit a significant low-frequency bias which would reveal an underlying ambiguity: empirically, the problem seems well-posed.

The objective of this paper is thus to establish the uniqueness of a solution to the problem of uncalibrated photometric stereo under general illumination, represented by first-order spherical harmonics. After discussing the classic case of directional lighting in Sect. 5.2, we characterize in Sect. 5.3 the ambiguities arising in uncalibrated photometric stereo under first-order spherical harmonics lighting. Then, we show in Sect. 5.4 that imposing integrability of the sought normal field resolves such ambiguities. In the orthographic case, only a global concave–convex ambiguity remains, hence the ambiguity is characterized by a single binary degree of freedom. For comparison, in the directional case, there are three real degrees of freedom characterizing the *generalized bas-relief* (GBR) ambiguity [4]. Moreover, under the perspective projection the problem becomes completely well-posed, which generalizes the result of [25] to more general lighting. In this case, the solution can even be determined in closed form, as shown in Sect. 5.5. Section 5.6 eventually recalls our findings, and suggests future research directions.

5.2   Preliminaries: Photometric Stereo Under Directional Lighting

Assuming a Lambertian surface is observed from a still camera under $m \geq 1$ different directional lighting indexed by $i \in \{1, \ldots, m\}$, the graylevel in the ith image can be modeled as follows:

$$I^i(x) = \rho(x)\, \mathbf{n}(x)^\top \mathbf{l}^i, \qquad \forall x \in \Omega, \tag{5.1}$$

where $\Omega \subset \mathbb{R}^2$ is the reconstruction domain (projection of the 3D-surface onto the image plane), $\rho(x) > 0$ is the albedo at the surface point conjugate to pixel x, $\mathbf{n}(x) \in \mathbb{S}^2 \subset \mathbb{R}^3$ is the unit-length outward normal at this point, and $\mathbf{l}^i \in \mathbb{R}^3$ is a vector oriented toward the light source, whose norm represents the relative intensity of the source. Photometric stereo consists, given a set of graylevel observations $I^i, i \in \{1, \ldots, m\}$, in estimating the shape (represented by the surface normal $\mathbf{n}$) and the reflectance (represented by the surface albedo ρ). Depending on whether the lighting vectors $\mathbf{l}^i$ are known or not, the problem is called calibrated or uncalibrated.

5.2.1   Calibrated Photometric Stereo Under Directional Lighting

Woodham showed in the late 70s [42] that $m \geq 3$ images captured under non-coplanar, known lighting vectors were sufficient to solve this problem. Indeed, defining for every $x \in \Omega$ the following observation vector $\mathbf{i}(x) \in \mathbb{R}^m$, lighting matrix $\mathbf{L} \in \mathbb{R}^{m \times 3}$ and surface vector $\mathbf{m}(x) \in \mathbb{R}^3$:

$$\mathbf{i}(x) = \begin{bmatrix} I^1(x) \\ \vdots \\ I^m(x) \end{bmatrix}, \qquad \mathbf{L} = \begin{bmatrix} \mathbf{l}^{1\top} \\ \vdots \\ \mathbf{l}^{m\top} \end{bmatrix}, \qquad \mathbf{m}(x) = \rho(x)\mathbf{n}(x), \tag{5.2}$$

the set of Eqs. (5.1) can be rewritten as a linear system in $\mathbf{m}(x)$:

$$\mathbf{i}(x) = \mathbf{L}\,\mathbf{m}(x), \qquad \forall x \in \Omega. \tag{5.3}$$

Provided that $\mathbf{L}$ is of rank three, (5.3) admits a unique least-squares solution in $\mathbf{m}(x)$, from which the normal and albedo can be extracted according to

$$\rho(x) = |\mathbf{m}(x)|, \qquad \mathbf{n}(x) = \frac{\mathbf{m}(x)}{|\mathbf{m}(x)|}. \tag{5.4}$$

Such a simple least-squares approach may be replaced by robust variational or learning-based strategies to ensure robustness [13, 17, 31, 34]. There also exist numerical solutions for handling non-Lambertian reflectance models [7, 15, 22, 40], non-distant light sources [18, 21, 30, 32], or the ill-posed cases where $m = 2$ [16, 20, 24, 29] or $m = 1$ [5, 8, 39, 45]. Such issues are not addressed in the present paper which rather focuses on the theoretical foundations of uncalibrated photometric stereo.

5.2.2 Uncalibrated Photometric Stereo Under Directional Lighting

The previous strategy relies on the knowledge of the lighting matrix $\mathbf{L}$, and it is not straightforward to extend it to unknown lighting. Let us illustrate this in the discrete setting, denoting the pixels as x_j, $j \in \{1, \ldots, n\}$ where $n = |\Omega|$ is the number of pixels, and stack all the observations in an observation matrix $\mathbf{I} \in \mathbb{R}^{m \times n}$ and all the surface vectors in a surface matrix $\mathbf{M} \in \mathbb{R}^{3 \times n}$:

$$\mathbf{I} = \big[\mathbf{i}(x_1), \ldots, \mathbf{i}(x_n)\big], \qquad \mathbf{M} = \big[\mathbf{m}(x_1), \ldots, \mathbf{m}(x_n)\big]. \tag{5.5}$$

Now, the set of n linear systems (5.3) can be represented compactly as:

$$\mathbf{I} = \mathbf{L}\mathbf{M} \tag{5.6}$$

where both the lighting matrix $\mathbf{L}$ and the surface matrix $\mathbf{M}$ are unknown. Since we know that $\mathbf{L}$ should be of rank three, a joint least-squares solution in $(\mathbf{L}, \mathbf{M})$ can be computed using truncated singular value decomposition [12]. Nevertheless, such a solution is not unique, since given a possible solution $(\mathbf{L}, \mathbf{M})$, any couple $(\mathbf{L}\mathbf{A}^{-1}, \mathbf{A}\mathbf{M})$ with $\mathbf{A} \in GL(3, \mathbb{R})$ is another solution:

$$\mathbf{I} = \mathbf{L}\mathbf{M} = \big(\mathbf{L}\mathbf{A}^{-1}\big)(\mathbf{A}\mathbf{M}), \qquad \forall \mathbf{A} \in GL(3, \mathbb{R}), \tag{5.7}$$

or equivalently, in the continuous setting:

$$\mathbf{i}(x) = \mathbf{L}\mathbf{m}(x) = \big(\mathbf{L}\mathbf{A}^{-1}\big)(\mathbf{A}\mathbf{m}(x)), \qquad \forall(x, \mathbf{A}) \in \Omega \times GL(3, \mathbb{R}). \tag{5.8}$$

However, not any surface matrix $\mathbf{M}$ (or $\mathbf{m}$-field, in the continuous setting) is acceptable as a solution. Indeed, this encodes the geometry of the surface, through its normals. Assuming that the surface is regular, its normals should satisfy the so-called integrability (or zero-curl) constraint. This constraint permits to reduce the ambiguities of uncalibrated photometric stereo, as shown in the next two subsections.

5.2.3 Integrability Under Orthographic Projection

Let us assume orthographic projection and denote $\mathbf{n}(x) := [n_1(x), n_2(x), n_3(x)]^\top$ the surface normal at 3D-point conjugate to pixel x. Let us further represent the surface as a Monge patch, i.e., a differentiable mapping $X : \Omega \to \mathbb{R}^3$ of the form $X(x) = (x, z(x))$, where $z : \Omega \to \mathbb{R}$ is a depth map. Let us assume this map z is twice differentiable, and let $\nabla z(x) = [z_u(x), z_v(x)]^\top \in \mathbb{R}^2$ be its gradient in some orthonormal basis (u, v) of the image plane. The integrability constraint is essentially a particular form of Schwarz' theorem, which implies that

$$z_{uv} = z_{vu} \quad \text{over } \Omega. \tag{5.9}$$

From the definition $\mathbf{n}(x) = \dfrac{\left[z_u(x), z_v(x), -1\right]^\top}{\sqrt{z_u(x)^2 + z_v(x)^2 + 1}}$ of the surface normal, and since $\mathbf{m}(x) = \rho(x)\,\mathbf{n}(x)$, Eq. (5.9) can be rewritten as

$$\left(\frac{m_1}{m_3}\right)_v = \left(\frac{m_2}{m_3}\right)_u \quad \text{over } \Omega. \tag{5.10}$$

Now, let us assume that one has found an $\mathbf{m}$-field solution of the left-hand side of (5.8), which further satisfies the integrability constraint (5.10) (in the discrete setting, this can be achieved using matrix factorization [44] or convex optimization techniques [36]). It can be shown that not all transformations $\mathbf{A}$ in the right-hand side of (5.8) preserve this constraint. Indeed, the only ones which are acceptable are those of the generalized bas-relief group. Such matrices define a subgroup of $GL(3, \mathbb{R})$ under the matrix product, and have the following form [4]:

$$\mathbf{G} = \begin{pmatrix} \lambda & 0 & -\mu \\ 0 & \lambda & -\nu \\ 0 & 0 & 1 \end{pmatrix}, \quad \mathbf{G}^{-1} = \begin{pmatrix} 1 & 0 & \mu/\lambda \\ 0 & 1 & \nu/\lambda \\ 0 & 0 & 1 \end{pmatrix}, \quad (\lambda, \mu, \nu) \in \mathbb{R}^3 \text{ and } \lambda \neq 0. \tag{5.11}$$

The three parameters μ, ν and λ characterize the GBR ambiguity inherent to uncalibrated photometric stereo under directional illumination and orthographic viewing. Intuitively, they can be understood as follows: any set of photometric stereo images can be reproduced by scaling the surface shape (this is the role of λ), adding a plane to it (this is the role of μ and ν), and moving the lighting vectors accordingly. If one is given a prior on the distribution of albedo values, on that of lighting vectors, or on the surface shape, then the three parameters can be estimated, i.e., the ambiguity can be resolved. The literature on that particular topic is extremely dense, see e.g., [38] for an overview, [6] for a modern numerical solution based on deep learning, and [27] for an application to RGB-D sensing. As we shall prove later in Sect. 5.4.1, in the case of nondirectional lighting represented using first-order spherical harmonics, the ambiguity is much simpler since it comes down to a global concave/convex one.

5.2.4 Integrability Under Perspective Projection

To terminate this discussion on uncalibrated photometric stereo under directional lighting, let us discuss the case of perspective projection, which was shown to be well-posed by Papadhimitri and Favaro in [25]. In the following, $x = (u, v)$ denotes the pixel coordinates with respect to the principal point (which is the projection of the camera center onto the image plane) and $f > 0$ denotes the focal length. The surface is now represented as the set of 3D-points $z(x) [u/f, v/f, 1]^\top$. Now, let us examine the perspective counterpart of the orthographic integrability constraint (5.10).

It is easy to show (see, e.g., [33]) that the surface normal is now defined as

$$\mathbf{n}(x) = \frac{1}{\sqrt{f^2 |\nabla z(x)|^2 + \left(-z(x) - [u, v]^\top \nabla z(x)\right)^2}} \begin{bmatrix} f \, \nabla z(x) \\ -z(x) - [u, v]^\top \nabla z(x) \end{bmatrix}. \tag{5.12}$$

If we define the log depth map as:

$$\tilde{z} = \log(z), \tag{5.13}$$

and denote:

$$p = -\frac{n_1}{n_3}, \qquad\qquad q = -\frac{n_2}{n_3}, \tag{5.14}$$

$$\hat{p} = \frac{p}{f - up - vq}, \qquad\qquad \hat{q} = \frac{q}{f - up - vq}, \tag{5.15}$$

then it is straightforward to show that

$$\nabla \tilde{z} = [\hat{p}, \hat{q}]^\top, \tag{5.16}$$

and that Schwarz' theorem (5.9) can be equivalently rewritten in terms of the gradient of the log depth map:

$$\tilde{z}_{uv} = \tilde{z}_{vu}. \tag{5.17}$$

This equation can be equivalently rewritten in terms of the coefficients of $\mathbf{m} = \rho \, \mathbf{n}$, just as we obtained (5.10) for the orthographic case. This rewriting is given by the following proposition, whose proof can be found in Appendix 1:

Proposition 5.1 Let $\mathbf{m} = [m_1, m_2, m_3]^\top : \Omega \to \mathbb{R}^3$ a field defined as $\mathbf{m} := \rho \, \mathbf{n}$, with $\rho : \Omega \to \mathbb{R}$ an albedo map and $\mathbf{n} : \Omega \to \mathbb{S}^2 \subset \mathbb{R}^3$ a normal field. The normal field $\mathbf{n}$ is integrable iff the coefficients of $\mathbf{m}$ satisfy the following relationship over Ω:

$$u(m_1 m_{2u} - m_{1u} m_2) + v(m_1 m_{2v} - m_{1v} m_2)$$
$$+ f(m_1 m_{3v} - m_{1v} m_3) + f(m_{2u} m_3 - m_2 m_{3u}) = 0. \tag{5.18}$$

The integrability constraint (5.18) is slightly more complicated than the orthographic one (5.10). Yet, this slight difference is of major importance, because the set of linear transformations $\mathbf{A}$ in (5.8) which preserve this condition is restricted to the identity matrix [25]. This means under perspective projection and directional lighting the uncalibrated photometric stereo problem is well-posed. As we shall prove later in Sect. 5.4.2, such a result can actually be extended to more general lighting represented using first-order spherical harmonics. Let us now elaborate on such a modeling of general lighting, and characterize the ambiguities therein.

5.3 Characterizing the Ambiguities in Uncalibrated Photometric Stereo Under General Lighting

The image formation model (5.1) is a simplified model, corresponding to the presence of a single light source located at infinity. However, this assumption is difficult to ensure in real-world experiments, and it would be more convenient to have at hand an image formation model accounting for general lighting (to handle multiple light sources, ambient lighting, etc.).

5.3.1 Spherical Harmonics Approximation of General Lighting

The most general image formation model for Lambertian surfaces would integrate the incident lighting received from all directions $\mathbf{u}_l \in \mathbb{S}^2$:

$$I^i(x) = \rho(x) \int_{\mathbb{S}^2} s^i(x, \mathbf{u}_l) \max\{\mathbf{n}(x)^\top \mathbf{u}_l, 0\} \, d\mathbf{u}_l, \qquad \forall x \in \Omega, \qquad (5.19)$$

where we denote $s^i(x, \mathbf{u}_l) \in \mathbb{R}$ the intensity of the light source in direction $\mathbf{u}_l \in \mathbb{S}^2$ at the surface point conjugate to pixel x in the ith image. In (5.19), the max operator encodes self-shadows: it ensures that the amount of reflected light does not take negative values for surface elements not facing the light source.

Assuming a single light source illuminates the scene in the ith image, and neglecting self-shadows, then Eq. (5.19) obviously comes down to the simplified model (5.1). However, there exist other simplifications of the integral model (5.19), which allow to handle more general illumination. Namely, the spherical harmonics approximation which were introduced simultaneously in [2, 35]. In the present work, we focus on first-order spherical harmonics approximation, which is known to capture approximately 87% of general lighting [10]. Using this approximation, (5.19) simplifies to (see the aforementioned papers for technical details):

$$I^i(x) = \rho(x) \begin{bmatrix} 1 \\ \mathbf{n}(x) \end{bmatrix}^\top \mathbf{l}^i, \qquad \forall x \in \Omega, \tag{5.20}$$

with $\mathbf{l}^i \in \mathbb{R}^4$ a vector representing the general illumination in the ith image. Denoting $\mathbf{L} = \begin{bmatrix} \mathbf{l}^1, \dots, \mathbf{l}^m \end{bmatrix}^\top \in \mathbb{R}^{m \times 4}$ the general lighting matrix, System (5.20) can be rewritten in the same form as the directional one (5.6):

$$\mathbf{i}(x) = \mathbf{Lm}(x), \qquad \forall x \in \Omega, \tag{5.21}$$

$$\text{with} \quad \mathbf{m}(x) = \rho(x) \begin{pmatrix} 1 \\ \mathbf{n}(x) \end{pmatrix}. \tag{5.22}$$

5.3.2 Uncalibrated Photometric Stereo Under First-Order Spherical Harmonics Lighting

Uncalibrated photometric stereo under first-order spherical harmonics lighting comes down to solving the set of linear systems (5.21) in terms of both the general lighting matrix $\mathbf{L}$ and the $\mathbf{m}$-field (which encodes albedo and surface normals). In the directional case discussed previously, this was possible only up to an invertible linear transformation, as shown by (5.8). Despite appearing more complicated at first glance, the case of first-order spherical harmonics is actually slightly more favorable than the directional one: not all such linear transformations are acceptable, because they have to preserve the particular form of the $\mathbf{m}$-field, given in Eq. (5.22). That is to say, given one $\mathbf{m}$-field solution and another one $\mathbf{m}^* = \mathbf{Am}$ obtained by applying an invertible linear transformation $\mathbf{A} \in GL(4, \mathbb{R})$, the entries c_1, c_2, c_3, c_4 of $\mathbf{m}^*$ should respect the constraint $c_1^2 = c_2^2 + c_3^2 + c_4^2$ over Ω (cf. Eq. (5.22), remembering that each surface normal has unit length).

As discussed in [3], this means that ambiguities in uncalibrated photometric stereo under first-order spherical harmonics are characterized as follows:

$$\mathbf{i}(x) = \mathbf{Lm}(x) = \left(\mathbf{LA}^{-1} \right) \left(\mathbf{Am}(x) \right), \qquad \forall (x, \mathbf{A}) \in \Omega \times L_s, \tag{5.23}$$

where L_s is the space of scaled Lorentz transformations defined by

$$L_s = \{ s\mathbf{A} \mid s \in \mathbb{R} \backslash \{0\} \text{ and } \mathbf{A} \in L \}, \tag{5.24}$$

with L the Lorentz group [28] arising in Einstein's theory of special relativity [9]:

$$L = \{ \mathbf{A} \in GL(4, \mathbb{R}) \mid \forall \mathbf{x} \in \mathbb{R}^4, \ l(\mathbf{Ax}) = l(\mathbf{x}) \}, \tag{5.25}$$

$$\text{with} \quad l : (t, x, y, z) \mapsto x^2 + y^2 + z^2 - t^2. \tag{5.26}$$

In spite of the presence of the scaled Lorentz ambiguity in Eq. (5.23), several heuristical approaches to solve uncalibrated photometric stereo under general lighting have been proposed lately. Let us mention the approaches based on hemispherical embedding [1] and on equivalent directional lighting [23], which both deal with the case of orthographic projection, and the variational approach in [11] for that of perspective projection. The empirically observed stability of such implementations tends to indicate that the problem might be better-posed than it seems, as already conjectured in [3]. In order to prove this conjecture, we will show in Sect. 5.4 that not all scaled Lorentz transformations preserve the integrability of surface normals. To this end, we need to characterize algebraically a scaled Lorentz transformation.

5.3.3 Characterization of the Scaled Lorentz Transformation

We propose to characterize any ambiguity matrix $\mathbf{A} \in L_s$ in (5.23) by means of a scale factor $s \neq 0$ (one degree of freedom), a vector inside the unit $\mathbb{R}^3$-ball $\mathbf{v} \in B(\mathbf{0}, 1)$ (three degrees of freedom, where $B(\mathbf{0}, 1) = \{\mathbf{x} \in \mathbb{R}^3, |x| < 1\}$) and a 3D-rotation matrix $\mathbf{O} \in SO(3, \mathbb{R})$ (three degrees of freedom, hence a total of seven). More explicitly, any scaled Lorentz transformation can be characterized algebraically as follows:

Theorem 5.1 *For any scaled Lorentz transformation* $\mathbf{A} \in L_s$, *there exists a unique triple* $(s, \mathbf{v}, \mathbf{O}) \in \mathbb{R}\backslash\{0\} \times B(\mathbf{0}, 1) \times SO(3, \mathbb{R})$ *such that*

$$
\mathbf{A} = s \begin{pmatrix} \epsilon_1(\mathbf{A})\, \gamma & \epsilon_1(\mathbf{A})\, \gamma\, \mathbf{v}^\top \mathbf{O} \\ \epsilon_2(\mathbf{A})\, \gamma\, \mathbf{v} & \epsilon_2(\mathbf{A})\, (\mathbf{I}_3 + \frac{\gamma^2}{1+\gamma}\mathbf{v}\mathbf{v}^\top)\mathbf{O} \end{pmatrix}, \tag{5.27}
$$

with

$$
\gamma = \frac{1}{\sqrt{1 - |\mathbf{v}|^2}}, \tag{5.28}
$$

$$
\epsilon_1(\mathbf{A}) = \begin{cases} 1 & \text{if } P_o(\mathbf{A}), \\ -1 & \text{else,} \end{cases} \tag{5.29}
$$

$$
\epsilon_2(\mathbf{A}) = \begin{cases} -1 & \text{if } (P_p(\mathbf{A}) \wedge \overline{P_o(\mathbf{A})}) \vee (\overline{P_p(\mathbf{A})} \wedge P_o(\mathbf{A})), \\ 1 & \text{else,} \end{cases} \tag{5.30}
$$

and $P_p(\mathbf{A})$ *stands for "$\mathbf{A}$ is proper", * $P_o(\mathbf{A})$ *for "$\mathbf{A}$ is orthochronous",*

where we recall that a Lorentz matrix $\mathbf{A}$ is "proper" iff it preserves the orientation of the Minkowski spacetime, and it is "orthochronous" iff it preserves the direction of the time, i.e.,

$$\mathbf{A} \in L \text{ is proper} \iff \det(\mathbf{A}) > 0, \tag{5.31}$$

$$\mathbf{A} \in L \text{ is orthochronous} \iff \forall \mathbf{x} = [t, x, y, z]^{\top} \in \mathbb{R}^4, \operatorname{sign}(t) = \operatorname{sign}(t'),$$

$$\text{where } \mathbf{A}\mathbf{x} = [t', x', y', z']^{\top}. \tag{5.32}$$

The opposites are improper and non-orthochronous, and we note L_o^p, L_o^i, L_n^p and L_n^i the sets of Lorentz transformations which are, respectively, proper and orthochronous, improper and orthochronous, proper and non-orthochronous, and improper and non-orthochronous. The Lorentz group is the union of all these spaces, i.e., $L = L_o^p \cup L_o^i \cup L_n^p \cup L_n^i$.

Using Theorem 5.1 (whose proof can be found in Appendix 2) to characterize the underlying ambiguity of uncalibrated photometric stereo under general lighting, we are ready to prove that imposing integrability disambiguates the problem.

5.4 Integrability Disambiguates Uncalibrated Photometric Stereo Under General Lighting

As we have seen in the previous section, uncalibrated photometric stereo under general lighting is ill-posed without further constraints, since it is prone to a scaled Lorentz ambiguity, cf. Eq. (5.23). Now, let us prove that not all scaled Lorentz transformations preserve the integrability of the underlying normal field.

We shall assume through the next two subsections that the pictured surface is twice differentiable and non-degenerate, in a sense which will be clarified in Sect. 5.4.3. Then, the only acceptable Lorentz transformation is the one which globally exchanges concavities and convexities in the orthographic case, while it is the identity in the perspective case. That is to say, the orthographic case suffers only from a global concave/convex ambiguity, while the perspective one is well-posed.

5.4.1 Orthographic Case

First, let us prove that under orthographic projection and first-order spherical harmonics lighting, there are only two integrable solutions to uncalibrated photometric stereo, and they differ by a global concave/convex transformation.

To this end, we consider the genuine solution $\mathbf{m}(x)$ of (5.21) corresponding to a normal field $\mathbf{n}(x)$ and albedo map $\rho(x)$, and another possible solution $\mathbf{m}^*(x) = \mathbf{A}\mathbf{m}(x)$, $\mathbf{A} \in L_s$, with $(\rho^*(x), \mathbf{n}^*(x))$ the corresponding albedo map and surface normals. The pictured surface being twice differentiable, the genuine normal field $\mathbf{n}$ is integrable by construction. We establish in this subsection that if the other candidate normal field $\mathbf{n}^*$ is assumed integrable as well, then both the genuine and the alternative solutions differ according to

$$\begin{cases} \rho^*(x) = \alpha\,\rho_j(x) \\ n_1^*(x) = \lambda\,n_1(x) \\ n_2^*(x) = \lambda\,n_2(x) \\ n_3^*(x) = n_3(x) \end{cases} \quad \forall x \in \Omega, \qquad (5.33)$$

where $\alpha > 0$ and $\lambda \in \{-1, 1\}$. That is to say, all albedo values are globally scaled by the same factor α, while the sign of the first two components of all normal vectors are inverted, i.e., concavities are turned into convexities and vice-versa. The global scale on the albedo should not be considered as an issue, since such values are relative to the camera response function and the intensities of the light sources, and they can be manually scaled back in a post-processing step if needed. However, the residual global concave/convex ambiguity shows that shape inference remains ill-posed. Still, the ill-posedness is characterized by a single binary degree of freedom, which is to be compared with the three real degrees of freedom characterizing the GBR ambiguity arising in the case of directional lighting [4].

More formally, this result can be stated as the following theorem, which characterizes the scaled Lorentz transformations in (5.23) preserving the integrability of the underlying normal field:

Theorem 5.2 *Under orthographic projection, the only scaled Lorentz transformation $A \in L_s$ which preserves integrability of normals is the following one, where $\alpha > 0$ and $\lambda \in \{-1, 1\}$:*

$$A = \alpha \begin{bmatrix} 1 & 0 & 0 & 0 \\ 0 & \lambda & 0 & 0 \\ 0 & 0 & \lambda & 0 \\ 0 & 0 & 0 & 1 \end{bmatrix}. \qquad (5.34)$$

Proof Let $m : \Omega \to \mathbb{R}^4$ a field with the form of Eq. (5.22), and let ρ and n the corresponding albedo map and normal field, assumed integrable. The normal field n being integrable, $p = -\dfrac{n_1}{n_3}$ and $q = -\dfrac{n_2}{n_3}$ satisfy the integrability constraint $p_v = q_u$ over Ω. Denoting by (c_1, c_2, c_3, c_4) the four components of the field m, and using the expression (5.22) of m, this implies

$$\left(\frac{c_2}{c_4}\right)_v = \left(\frac{c_3}{c_4}\right)_u \quad \text{over } \Omega,$$

$$\iff \frac{c_{2v}c_4 - c_2 c_{4v}}{c_4{}^2} = \frac{c_{3u}c_4 - c_3 c_{4u}}{c_4{}^2} \quad \text{over } \Omega,$$

$$\iff (c_{2v} - c_{3u})c_4 + c_{4u}c_3 - c_{4v}c_2 = 0 \quad \text{over } \Omega. \qquad (5.35)$$

Let $m^* = Am$, with A a scaled Lorentz transformation having the form given by Theorem 5.1, and let ρ^* and n^* the corresponding albedo map and normal field, assumed integrable. The same rationale as above on the alternative normal field n^*

yields

$$(c_{2v}^* - c_{3u}^*)c_4^* + c_{4u}^* c_3^* - c_{4v}^* c_2^* = 0 \qquad \text{over } \Omega. \tag{5.36}$$

Since $\mathbf{m}^* = \mathbf{Am}$, (5.36) writes as

$$
\begin{aligned}
(A_{21}c_{1v} + A_{22}c_{2v} &+ A_{23}c_{3v} + A_{24}c_{4v} - A_{31}c_{1u} - A_{32}c_{2u} \\
&- A_{33}c_{3u} - A_{34}c_{4u})(A_{41}c_1 + A_{42}c_2 + A_{43}c_3 + A_{44}c_4) \\
+ (A_{41}c_{1u} + A_{42}c_{2u} &+ A_{43}c_{3u} + A_{44}c_{4u})(A_{31}c_1 + A_{32}c_2 + A_{33}c_3 + A_{34}c_4) \\
- (A_{41}c_{1v} + A_{42}c_{2v} &+ A_{43}c_{3v} + A_{44}c_{4v})(A_{21}c_1 + A_{22}c_2 + A_{23}c_3 + A_{24}c_4) \\
&= 0 \quad \text{over } \Omega. \tag{5.37}
\end{aligned}
$$

Let us introduce the following notation, $1 \leq i < j \leq 4$, and $k \in \{u, v\}$:

$$c_k^{i,j}(x) = c_j(x)c_{ik}(x) - c_i(x)c_{jk}(x), \qquad \forall x \in \Omega, \tag{5.38}$$

and denote as follows the minors of size two of matrix $\mathbf{A}$:

$$A_{k,l}^{i,j} = A_{ij}A_{kl} - A_{kj}A_{il}, \qquad 1 \leq i < k \leq 4, \; 1 \leq j < l \leq 4. \tag{5.39}$$

Then, factoring (5.37) firstly by the coefficients A_{ij} and after by $c_u^{i,j}$ and $c_v^{i,j}$ for every $(i, j) \in \{1, 2, 3, 4\}$ with $i < j$, we get

$$
\begin{aligned}
c_v^{1,2} A_{4,2}^{2,1} + c_v^{1,3} A_{4,3}^{2,1} &+ c_v^{1,4} A_{4,4}^{2,1} + c_v^{2,3} A_{4,3}^{2,2} + c_v^{2,4} A_{4,4}^{2,2} + c_v^{3,4} A_{4,4}^{2,3} \\
- c_u^{1,2} A_{4,2}^{3,1} - c_u^{1,3} A_{4,3}^{3,1} &- c_u^{1,4} A_{4,4}^{3,1} - c_u^{2,3} A_{4,3}^{3,2} - c_u^{2,4} A_{4,4}^{3,2} - c_u^{3,4} A_{4,4}^{3,3} = 0 \text{ over } \Omega. \tag{5.40}
\end{aligned}
$$

In addition, (5.35) also writes as

$$c_v^{2,4} = c_u^{3,4} \qquad \text{over } \Omega. \tag{5.41}$$

Thus, substituting $c_v^{2,4}$ by $c_u^{3,4}$, Eq. (5.40) can be rewritten as

$$\mathbf{i}_o(x)^\top \mathbf{a} = 0, \qquad \forall x \in \Omega, \tag{5.42}$$

where $\mathbf{i}_o(x) \in \mathbb{R}^{11}$ is the "orthographic integrability vector" containing factors $c_u^{i,j}(x)$ and $c_v^{i,j}(x)$, and $\mathbf{a} \in \mathbb{R}^{11}$ contain the minors $A_{k,l}^{i,j}$ of $\mathbf{A}$ appearing in (5.40).

Since the surface is assumed to be non-degenerate (cf. Sect. 5.4.3), there exist at least 11 points $x \in \Omega$ such that a full-rank matrix can be formed by concatenating the vectors $\mathbf{i}_o(x)^\top$ row-wise. We deduce that the only solution to (5.42) is $\mathbf{a} = 0$, which is equivalent to the following equations:

$$\begin{cases} A_{4,3}^{3,2} = A_{4,4}^{3,2} = A_{4,4}^{2,3} = A_{4,3}^{2,2} = 0, \\ A_{4,4}^{3,3} = A_{4,4}^{2,2}, \\ A_{4,2}^{2,1} = A_{4,2}^{3,1} = A_{4,3}^{2,1} = A_{4,3}^{3,1} = A_{4,4}^{2,1} = A_{4,4}^{3,1} = 0. \end{cases} \tag{5.43}$$

According to Corollary 5.1 provided in Appendix 3, this implies that the submatrix of $\mathbf{A}$ formed by its last three rows and columns is a scaled generalized bas-relief transformation, i.e., there exists a unique quadruple $(\lambda, \mu, \nu, \beta) \in \mathbb{R}^4$ with $\lambda \neq 0$, $\beta \neq 0$, such that

$$\mathbf{A} = \begin{pmatrix} A_{11} & A_{12} & A_{13} & A_{14} \\ A_{21} & \beta\lambda & 0 & -\beta\mu \\ A_{31} & 0 & \beta\lambda & -\beta\nu \\ A_{41} & 0 & 0 & \beta \end{pmatrix}. \tag{5.44}$$

By taking into account the last equation of System (5.43), we get

$$\mathbf{A} = \begin{pmatrix} A_{11} & A_{12} & A_{13} & A_{14} \\ 0 & \beta\lambda & 0 & -\beta\mu \\ 0 & 0 & \beta\lambda & -\beta\nu \\ 0 & 0 & 0 & \beta \end{pmatrix}. \tag{5.45}$$

Identifying (5.45) with the expression in Theorem 5.1, $\mathbf{v} = \mathbf{0}$, $\gamma = 1$ and $s \, \epsilon_2(\mathbf{A})$
$\mathbf{O} = \beta \begin{pmatrix} \lambda & 0 & -\mu \\ 0 & \lambda & -\nu \\ 0 & 0 & 1 \end{pmatrix}$. In addition, $\mathbf{O} \in SO(3, \mathbb{R})$, which implies $\mathbf{O}^\top \mathbf{O} = \mathbf{I_3}$. Thus,
since $\epsilon_2(\mathbf{A})^2 = 1$, we have $(s \, \epsilon_2(\mathbf{A}) \, \mathbf{O})^\top \epsilon_2(\mathbf{A}) \, \mathbf{O}) = s^2 \, \mathbf{I_3}$. Equivalently,

$$\beta \begin{pmatrix} \lambda & 0 & 0 \\ 0 & \lambda & 0 \\ -\mu & -\nu & 1 \end{pmatrix} \beta \begin{pmatrix} \lambda & 0 & -\mu \\ 0 & \lambda & -\nu \\ 0 & 0 & 1 \end{pmatrix} = \begin{pmatrix} s^2 & 0 & 0 \\ 0 & s^2 & 0 \\ 0 & 0 & s^2 \end{pmatrix},$$
$$\Longleftrightarrow \beta^2 \begin{pmatrix} \lambda^2 & 0 & -\mu\lambda \\ 0 & \lambda^2 & -\nu\lambda \\ -\mu\lambda & -\nu\lambda & \mu^2 + \nu^2 + 1 \end{pmatrix} = \begin{pmatrix} s^2 & 0 & 0 \\ 0 & s^2 & 0 \\ 0 & 0 & s^2 \end{pmatrix}, \tag{5.46}$$

which implies $\lambda^2 = 1$, $\mu = 0$, $\nu = 0$, $\beta^2 = s^2$.

Finally, $\det(s \, \epsilon_2(\mathbf{A}) \, \mathbf{O}) = \beta \, \lambda^2 \, \beta = \epsilon_2(\mathbf{A}) \, s$, thus according to (5.45)

$$\mathbf{A} = s \begin{pmatrix} \epsilon_1(\mathbf{A}) & 0 & 0 & 0 \\ 0 & \epsilon_2(\mathbf{A})\lambda & 0 & 0 \\ 0 & 0 & \epsilon_2(\mathbf{A})\lambda & 0 \\ 0 & 0 & 0 & \epsilon_2(\mathbf{A}) \end{pmatrix}. \tag{5.47}$$

Plugging (5.47) into $\mathbf{m}^* = \mathbf{Am}$, we obtain

$$\begin{pmatrix} \rho^*(x) \\ \rho^*(x)\, n_1^*(x) \\ \rho^*(x)\, n_2^*(x) \\ \rho^*(x)\, n_3^*(x) \end{pmatrix} = s \begin{pmatrix} \epsilon_1(\mathbf{A})\, \rho(x) \\ \epsilon_2(\mathbf{A})\, \lambda\, \rho(x)\, n_1(x) \\ \epsilon_2(\mathbf{A})\, \lambda\, \rho(x)\, n_2(x) \\ \epsilon_2(\mathbf{A})\, \rho(x)\, n_3(x) \end{pmatrix}, \qquad \forall x \in \Omega. \tag{5.48}$$

Now, knowing that albedos $\rho, \rho^* \geqslant 0$ (they represent the proportion of light which is reflected by the surface), and that the last component of normals $n_3, n_3^* \leq 0$ (the normals point toward the camera), Eq. (5.48) implies that $\epsilon_1(\mathbf{A})$ and $\epsilon_2(\mathbf{A})$ have exactly the same sign as s.

Two cases must be considered. If $s > 0$, then $\epsilon_1(\mathbf{A}) = \epsilon_2(\mathbf{A}) = 1$, and plugging these values into (5.47) we obtain the expression provided in Theorem 5.2. If $s < 0$, then $\epsilon_1(\mathbf{A}) = \epsilon_2(\mathbf{A}) = -1$, and we again get the expression provided in Theorem 5.2. □

From a practical point of view, once an integrable normal field candidate has been found heuristically, using, e.g., hemispherical embedding [1] or an equivalent directional lighting model [23], the residual ambiguity, i.e., the sign of λ, needs to be set manually, as proposed for instance in [23]. As we shall see now, in the case of perspective projection the problem becomes even completely well-posed, which circumvents the need for any manual intervention.

5.4.2 Perspective Case

Now we will prove that uncalibrated photometric stereo under first-order spherical harmonics lighting and perspective projection is well-posed. This means, imposing integrability restricts the admissible ambiguity matrices $\mathbf{A}$ in (5.23) to the identity matrix (up to a factor scaling all albedo values without affecting the geometry):

Theorem 5.3 *Under perspective projection, the only scaled Lorentz transformation* $\mathbf{A} \in L_s$ *which preserves integrability of normals is the identity matrix, up to a scale factor* $\alpha > 0$:

$$\mathbf{A} = \alpha \mathbf{I}_4. \tag{5.49}$$

Proof Let $\mathbf{m} : \Omega \to \mathbb{R}^4$ a field with the form of Eq. (5.22), whose normal field is integrable. Let $\mathbf{m}^* = \mathbf{A}\mathbf{m}$ another such field whose normal field is integrable, with $\mathbf{A} \in L_s$ a scaled Lorentz transformation having the form given by Theorem 5.1.

Let us denote by (c_1, c_2, c_3, c_4) the four components of the field $\mathbf{m}$, and by $(c_1^*, c_2^*, c_3^*, c_4^*)$ those of $\mathbf{m}^*$. According to Proposition 5.1, the integrability constraint of the normal field associated with $\mathbf{m}^*$ writes as follows:

$$u(c^*)_u^{2,3} + v(c^*)_v^{2,3} + f(c^*)_v^{2,4} - f(c^*)_u^{3,4} = 0 \qquad \text{over } \Omega, \tag{5.50}$$

with the same notations as in (5.38).

As in the previous proof of Theorem 5.2, we substitute in the integrability constraint (5.50) the entries of $\mathbf{m}^* = \mathbf{A}\mathbf{m}$ with their expressions in terms of entries of $\mathbf{A}$ and $\mathbf{m}$. Then, by factoring firstly by the coefficients A_{ij} and then by $c_u^{i,j}$ and $c_v^{i,j}$ for every $(i, j) \in \{1, 2, 3, 4\}$ with $i < j$, we get

$$
\begin{aligned}
& \left(uc_u^{1,2} + vc_v^{1,2}\right) A_{3,2}^{2,1} + \left(uc_u^{1,3} + vc_v^{1,3}\right) A_{3,3}^{2,1} \\
& + \left(uc_u^{1,4} + vc_v^{1,4}\right) A_{3,4}^{2,1} + \left(uc_u^{2,3} + vc_v^{2,3}\right) A_{3,3}^{2,2} \\
& + \left(uc_u^{2,4} + vc_v^{2,4}\right) A_{3,4}^{2,2} + \left(uc_u^{3,4} + vc_v^{3,4}\right) A_{3,4}^{2,3} \\
& + f \left(c_v^{1,2} A_{4,2}^{2,1} + c_v^{1,3} A_{4,3}^{2,1} + c_v^{1,4} A_{4,4}^{2,1} + c_v^{2,3} A_{4,3}^{2,2} + c_v^{2,4} A_{4,4}^{2,2} + c_v^{3,4} A_{4,4}^{2,3}\right) \\
& - f \left(c_u^{1,2} A_{4,2}^{3,1} + c_u^{1,3} A_{4,3}^{3,1} + c_u^{1,4} A_{4,4}^{3,1} + c_u^{2,3} A_{4,3}^{3,2} + c_u^{2,4} A_{4,4}^{3,2} + c_u^{3,4} A_{4,4}^{3,3}\right) \\
& = 0 \qquad \text{over } \Omega.
\end{aligned}
\tag{5.51}
$$

By concatenating Eqs. (5.51) for all pixels $x \in \Omega$, we get the following set of linear systems:
$$
\mathbf{i}_p(x)^\top \mathbf{w} = 0, \qquad \forall x \in \Omega,
\tag{5.52}
$$

where $\mathbf{w} \in \mathbb{R}^{18}$ contains all the minors $A_{k,l}^{i,j}$ of the ambiguity matrix $\mathbf{A}$ in Eq. (5.51), and the "perspective integrability" vector $\mathbf{i}_p(x)$ depends only on u, v, f, and $c_k^{i,j}$, i.e., known quantities. We will see later in Sect. 5.5, that numerically solving the set of Eqs. (5.52) provides a simple way to numerically solve uncalibrated perspective photometric stereo under first-order spherical harmonics lighting.

If in addition we use the fact that $\mathbf{m}$ fulfills the integrability constraint (5.50), we can substitute $(uc_u^{2,3} + vc_v^{2,3})$ by $(-fc_v^{2,4} + fc_u^{3,4})$ in Eq. (5.51), and we get 17 summands instead of 18, turning (5.52) as follows:

$$
\mathbf{c}(x)^\top \mathbf{a} = 0 \qquad \text{over } \Omega,
\tag{5.53}
$$

where $\mathbf{c}(x), \mathbf{a} \in \mathbb{R}^{17}$.

Since the surface is assumed to be non-degenerate (cf. Sect. 5.4.3), there exist at least 17 points $x \in \Omega$ such that a full-rank matrix can be formed by row-wise concatenation of vectors $\mathbf{c}(x)^\top$, $x \in \Omega$. We deduce that $\mathbf{a} = 0$ and we get the following equations:

$$
\begin{cases}
A_{4,3}^{3,2} = A_{4,4}^{3,2} = A_{4,4}^{2,3} = A_{4,3}^{2,2} = 0, \\
A_{3,3}^{2,2} = A_{4,4}^{3,3}, \\
A_{3,3}^{2,2} = A_{4,4}^{2,2}, \\
A_{3,2}^{2,1} = A_{3,3}^{2,1} = A_{3,4}^{2,1} = A_{3,4}^{2,2} = A_{3,4}^{2,3} = 0, \\
A_{4,2}^{2,1} = A_{4,3}^{2,1} = A_{4,4}^{2,1} = A_{4,3}^{2,2} = A_{4,4}^{2,3} = 0, \\
A_{4,3}^{3,2} = A_{4,4}^{3,2} = A_{4,2}^{3,1} = A_{4,3}^{3,1} = A_{4,4}^{3,1} = 0.
\end{cases}
\tag{5.54}
$$

According to the first three equations of System (5.54), the submatrix of $\mathbf{A}$ formed by the last three rows and columns is a scaled generalized bas-relief transformation (see Corollary 5.1 in Appendix 3). That is to say, there exists a unique quadruple $(\lambda, \mu, \nu, \alpha) \in \mathbb{R}^4$ with $\lambda \neq 0, \alpha \neq 0$, such that

$$
\mathbf{A} = \begin{pmatrix} A_{11} & A_{12} & A_{13} & A_{14} \\ A_{21} & \alpha\lambda & 0 & -\alpha\mu \\ A_{31} & 0 & \alpha\lambda & -\alpha\nu \\ A_{41} & 0 & 0 & \alpha \end{pmatrix}. \tag{5.55}
$$

Taking into account the other equations of system (5.54), we get $\lambda = 1, \mu = \nu = A_{21} = A_{31} = A_{41} = 0$. Then, the same arguments as those used around Eq. (5.47) yield $A_{12} = A_{13} = A_{14} = 0$, and the form (5.22) of $\mathbf{m}^* = \mathbf{Am}$ implies $A_{11} = \alpha$, which concludes the proof.   □

Let us remark that such a particular form of a scaled Lorentz transformation only scales all albedo values, leaving the geometry unchanged. From a practical point of view, this means that once an integrable candidate has been found, it corresponds to the genuine surface and there is no need to manually solve any ambiguity, unlike in the orthographic case. In Sect. 5.5, we will see that such a candidate can be estimated in closed form in the discrete setting. This will allow us to empirically verify the validity of our theoretical results, through numerical experiments on simulated images. Before that, let us briefly elaborate on degenerate surfaces, i.e., surfaces for which the two theorems in the present section do not hold.

5.4.3 Degenerate Surfaces

The two previous theorems rely on the assumption that the surface is non-"degenerate". Although degenerate surfaces are rarely encountered in practice, this notion needs to be clarified for the completeness of this study.

Degenerate surfaces are those having a particularly simple shape, which causes the matrix formed by the concatenation of the integrability vectors ($\mathbf{i}_o(x)$ in the orthographic case, cf. (5.42), or $\mathbf{c}(x)$ in the perspective case, cf. (5.53)) not to be full-rank. Here we algebraically characterize such surfaces, for which the integrability constraint is not enough to solve the Lorentz ambiguity.

5.4.3.1 Orthographic Case

Let $\mathbf{m} : \Omega \rightarrow \mathbb{R}^4$ be a field of the form of (5.22), and let ρ and $\mathbf{n}$ be the corresponding albedo map and normal field, respectively. We denote by (c_1, c_2, c_3, c_4) the four components of the field $\mathbf{m}$, and use the definition (5.38) of the coefficients $c_k^{i,j}$, $i, j \in \{1, \ldots, 4\}, k \in \{u, v\}$. Then, the surface defined by the field $\mathbf{m}$ is degenerate iff the $\left(c_k^{i,j}\right)_{(k,i,j)\neq(v,2,4)}$ are linearly dependent, i.e., if there exists a nonzero vector

$(\lambda_k^{i,j}) \in \mathbb{R}^{11}\backslash\{0\}$ such that for any pixel $x \in \Omega$

$$\sum_{\substack{k \in \{u,v\} \\ 1 \leq i < j \leq 4 \\ (k,i,j) \neq (v,2,4)}} \lambda_k^{i,j} c_k^{i,j}(x) = 0. \tag{5.56}$$

To illustrate this notion on some examples, let us remark that by definition of the coefficients $\left(c_k^{i,j}\right)$:

$$\rho\,\mathbf{n} \times \rho\,\mathbf{n}_u = \begin{pmatrix} -c_u^{3,4} \\ c_u^{2,4} \\ -c_u^{2,3} \end{pmatrix}, \qquad \rho\,\mathbf{n} \times \rho\,\mathbf{n}_v = \begin{pmatrix} -c_v^{3,4} \\ c_v^{2,4} \\ -c_v^{2,3} \end{pmatrix} \qquad \text{over } \Omega, \tag{5.57}$$

where $\times$ denotes the cross-product.

Therefore, the following sufficient (but not necessary) conditions to be a degenerate surface can be formulated:

- $\mathbf{n}_u = \mathbf{n}_v = 0$: a planar surface.
- $\mathbf{n}_u = 0$ and $\mathbf{n}_v \neq 0$: a surface with vanishing curvature along u (see Fig. 5.1a);
- $\mathbf{n}_u \neq 0$ and $\mathbf{n}_v = 0$: a surface with vanishing curvature along v (see Fig. 5.1b);
- $\mathbf{n}_u = \mathbf{n}_v$: a surface with vanishing curvature along $u = -v$ (see Fig. 5.1c);
- $\mathbf{n}_u = -\mathbf{n}_v$, a surface with vanishing curvature along $u = v$ (see Fig. 5.1d).

5.4.3.2 Perspective Case

Analogously, a surface is degenerate under perspective projection iff there exists a nonzero vector $\left((\alpha_k^{i,j}), (\beta_{i,j})_{(i,j)\neq(2,3)}\right) \in \mathbb{R}^{17}\backslash\{0\}$ such that, for any pixel $x = (u, v) \in \Omega$:

$$\left[\sum_{k \in \{u,v\}} \sum_{1 \leq i < j \leq 4} \left(\alpha_k^{i,j}\right) f c_k^{i,j}(x)\right] + \sum_{\substack{1 \leq i < j \leq 4 \\ (i,j) \neq (2,3)}} \beta_{i,j} \left(u c_u^{i,j}(x) + v c_v^{i,j}(x)\right) = 0,$$

$$\tag{5.58}$$

where f is the focal length.

The surfaces shown in Fig. 5.1 are examples of degenerate surfaces. There exist other examples, yet it is not straightforward to characterize them in a simple way. On the other hand, in practice if the surface was simple enough to yield degeneracy, one would not resort to photometric stereo at all. In real-world problems, the geometry of the pictured surface is rich enough, so degenerate surfaces rarely or never arise. This means, it is possible to numerically solve equations such as (5.52) in a stable manner. As shown in the next section, this provides a practical way to numerically solve perspective uncalibrated photometric stereo under general lighting.

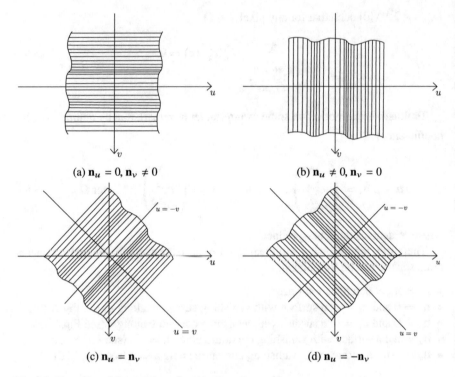

(a) $\mathbf{n}_u = 0, \mathbf{n}_v \neq 0$ (b) $\mathbf{n}_u \neq 0, \mathbf{n}_v = 0$

(c) $\mathbf{n}_u = \mathbf{n}_v$ (d) $\mathbf{n}_u = -\mathbf{n}_v$

Fig. 5.1 Examples of degenerate surfaces in the orthographic case

5.5 Numerical Solving of the Perspective Case

In this section, we derive a practical algorithm for solving perspective uncalibrated photometric stereo under first-order spherical harmonics lighting. More specifically, we provide a closed-form solution for an integrable normal field satisfying the image formation model (5.21), provided that the perspective camera is calibrated (i.e., its focal length and principal point are known).

5.5.1 Discrete Formulation

First, let us reformulate the problem in the discrete setting. Let us stack all the observations $I^i(x)$, $i \in \{1, \ldots, m\}$, $x \in \Omega$, in a matrix $\mathbf{I} \in \mathbb{R}^{m \times n}$, with $n = |\Omega|$ the number of pixels. Similarly to the directional lighting case represented by (5.6), the set of linear systems (5.21) can be rewritten in matrix form as

$$\mathbf{I} = \mathbf{LM}, \tag{5.59}$$

where $\mathbf{L} \in \mathbb{R}^{m \times 4}$ is the general lighting matrix, and $\mathbf{M} \in \mathbb{R}^{4 \times n}$ stacks all the unknown $\mathbf{m}(x)$-vectors columnwise (each column $\mathbf{m}_j = \mathbf{m}(x_j)$ has thus the form given in Eq. (5.22)).

As shown in [3], a least-squares solution $(\mathbf{L}_1, \mathbf{M}_1)$ of (5.59) satisfying the constraint (5.22) can be obtained by singular value decomposition of $\mathbf{I}$. Since we know that any other $\mathbf{M}$-matrix solution of (5.59) differs from $\mathbf{M}_1$ according to a scaled Lorentz transform, the genuine solution $\mathbf{M}^* \in \mathbb{R}^{4 \times n}$ is given by

$$\mathbf{M}^* = \mathbf{A}\mathbf{M}_1, \tag{5.60}$$

with $\mathbf{A} \in L_s$ an unknown scaled Lorentz transformation.

In the last section, we have seen that there exists a unique $\mathbf{m}$-field which both satisfies the image formation model and is integrable. This means, that if the pictured surface is twice differentiable and non-degenerate, then matrix $\mathbf{A}$ in (5.60) is unique (up to scale). In fact, we only need the last three rows of matrix $\mathbf{A}$: left-multiplying the last three rows of the initial guess $\mathbf{M}_1$ by this submatrix, we obtain a matrix of size $3 \times n$ where the norm of the jth column is the albedo at the surface point conjugate to pixel x_j, and normalizing each column yields the surface normal at this point.

The problem thus comes down to estimating the last three rows of matrix $\mathbf{A}$. According to Proposition 5.8 in Appendix 3, these rows can be written in the form $(\mathbf{v} \mid \mathbf{Q}) \in \mathbb{R}^{3 \times 4}$, where $\mathbf{v} \in \mathbb{R}^3$ and $\mathbf{Q} \in GL(3, \mathbb{R})$. Next we show how to estimate $\mathbf{v}$ and $\mathbf{Q}$ in closed form, using a discrete analogous of the perspective integrability constraint.

5.5.2 Closed-Form Solution Through Discrete Integrability Enforcement

During the proof of Theorem 5.3, we showed that the integrability constraint yields the set of linear systems (5.52) over Ω. In the discrete setting, this set of equations can be written compactly as

$$\mathbf{I}_p \mathbf{w} = 0, \tag{5.61}$$

where $\mathbf{w} \in \mathbb{R}^{18}$ contains several minors of size 2 denoted by $\left(A_{k,l}^{i,j} \right)$, and $\mathbf{I}_p \in \mathbb{R}^{n \times 18}$ is a "perspective integrability matrix" depending only upon the known camera parameters and entries of $\mathbf{M}_1$.

Matrix $\mathbf{I}_p$ is in general full-rank. Thus, the least-squares solution (up to scale) of (5.61) in terms of vector $\mathbf{w}$ can be determined by singular value decomposition of $\mathbf{I_p}$: denoting by $\mathbf{I_p} = \mathbf{U}\boldsymbol{\Sigma}\mathbf{V}^\top$ this decomposition, the solution $\mathbf{w}$ is the last column of $\mathbf{V}$. We denote by $\left(\tilde{A}_{k,l}^{i,j} \right) = \left(\lambda \, A_{k,l}^{i,j} \right)$ its entries, where $\lambda \neq 0$ denotes the unknown scale factor.

Now, recall that matrix $\mathbf{Q} \in \mathbb{R}^{3 \times 3}$ to be determined is the sub-matrix formed by the last three rows and columns of $\mathbf{A}$. It relates to the aforementioned minors according to

$$\mathbf{Q}^{-1} = \frac{1}{\det(\mathbf{Q})} \operatorname{com}(\mathbf{Q})^{\top} = \frac{1}{\det(\mathbf{Q})} \begin{pmatrix} A_{4,4}^{3,3} & -A_{4,4}^{2,3} & A_{3,4}^{2,3} \\ -A_{4,4}^{3,2} & A_{4,4}^{2,2} & -A_{3,4}^{2,2} \\ A_{4,3}^{3,2} & -A_{4,3}^{2,2} & A_{3,3}^{2,2} \end{pmatrix}, \tag{5.62}$$

where $\operatorname{com}(\mathbf{Q})$ is the comatrix of $\mathbf{Q}$. Thus,

$$\lambda \mathbf{Q}^{-1} = \frac{1}{\det(\mathbf{Q})} \mathbf{\Delta}^{-1}, \text{ where } \mathbf{\Delta} = \begin{pmatrix} \tilde{A}_{4,4}^{3,3} & -\tilde{A}_{4,4}^{2,3} & \tilde{A}_{3,4}^{2,3} \\ -\tilde{A}_{4,4}^{3,2} & \tilde{A}_{4,4}^{2,2} & -\tilde{A}_{3,4}^{2,2} \\ \tilde{A}_{4,3}^{3,2} & -\tilde{A}_{4,3}^{2,2} & \tilde{A}_{3,3}^{2,2} \end{pmatrix}^{-1}. \tag{5.63}$$

Hence, we can determine $\mathbf{Q}$ up to scale:

$$\mathbf{Q} = (\lambda \det\mathbf{Q}) \, \mathbf{\Delta}. \tag{5.64}$$

Next, we turn our attention to the estimation of vector $\mathbf{v} \in \mathbb{R}^3$ (recall that this vector is formed by the first column and last three rows of $\mathbf{A}$), up to scale. To this end, we consider the last nine minors. For example, considering $\tilde{A}_{3,2}^{2,1}$:

$$\tilde{A}_{3,2}^{2,1} = \lambda \, (A_{21} A_{32} - A_{31} A_{22}) \tag{5.65}$$

$$= \lambda \, (A_{21} Q_{21} - A_{31} Q_{11}) \tag{5.66}$$

$$\underset{(5.64)}{=} \left(\lambda^2 \det(\mathbf{Q}) A_{21}\right) \Delta_{21} - \left(\lambda^2 \det(\mathbf{Q}) A_{31}\right) \Delta_{11}. \tag{5.67}$$

Let $\hat{\mathbf{v}} = \begin{pmatrix} \hat{v}_1 \\ \hat{v}_2 \\ \hat{v}_3 \end{pmatrix} = \lambda^2 \det(\mathbf{Q}) \begin{pmatrix} A_{21} \\ A_{31} \\ A_{41} \end{pmatrix} = (\lambda^2 \det(\mathbf{Q}))\mathbf{v}$. Equation (5.67) can be written as:

$$\Delta_{21} \hat{v}_1 - \Delta_{11} \hat{v}_2 = \hat{A}_{3,2}^{2,1}. \tag{5.68}$$

In the same manner, by using all the other minors which involve A_{21}, A_{31} or A_{41}, we get the following over-constrained linear system:

$$\mathbf{S}\hat{\mathbf{v}} = \mathbf{b}, \tag{5.69}$$

where $\mathbf{S} \in \mathbb{R}^{9 \times 3}$ and $\mathbf{b} \in \mathbb{R}^9$. A least-squares solution for $\hat{\mathbf{v}}$ can be found using, e.g., the pseudo inverse:

$$\hat{\mathbf{v}} = \mathbf{S}^{\dagger}\mathbf{b}. \tag{5.70}$$

Besides,

$$\lambda^2 \det(\mathbf{Q})(\mathbf{v} \mid \mathbf{Q}) = (\lambda^2 \det(\mathbf{Q})\mathbf{v} \mid \lambda^2 \det(\mathbf{Q})\mathbf{Q}) \tag{5.71}$$

$$= (\hat{\mathbf{v}} \mid \lambda^3 \det(\mathbf{Q})^2 \mathbf{\Delta}), \tag{5.72}$$

and applying the determinant to both sides of Eq. (5.64) yields

$$\lambda^3 \det(\mathbf{Q})^2 = \frac{1}{\det(\mathbf{\Delta})}. \tag{5.73}$$

Plugging (5.73) into (5.72), we eventually obtain the following closed-form expression for $(\mathbf{v} \mid \mathbf{Q})$:

$$(\mathbf{v} \mid \mathbf{Q}) = \frac{1}{\lambda^2 \det(\mathbf{Q})} \left(\hat{\mathbf{v}} \mid \frac{1}{\det(\mathbf{\Delta})} \mathbf{\Delta} \right). \tag{5.74}$$

Since λ and $\det(\mathbf{Q})$ in (5.74) are unknown, the solution $(\mathbf{v} \mid \mathbf{Q})$ is known only up to scale. As already stated, the actual value of this scale factor is not important, since it only scales all albedo values simultaneously without affecting the geometry. Let us denote by $\tilde{\mathbf{M}}_1$ the submatrix formed by the last three rows of the initial guess $\mathbf{M}_1$. Then, matrix $\tilde{\mathbf{M}}_2 = (\mathbf{v} \mid \mathbf{Q})\tilde{\mathbf{M}}_1$ is a $3 \times n$ matrix where each column corresponds to one surface normal, scaled by the albedo. The norm of each column of $\tilde{\mathbf{M}}_2$ thus provides the sought albedo (up to scale), and normalizing each column provides the sought surface normal.

Therefore, we now have at hand a practical way to find an integrable normal field solving uncalibrated photometric stereo under general lighting and perspective projection. In the next subsection, we show on simulated data that such a solution indeed corresponds to the genuine surface, which provides an empirical evidence for the theoretical analysis conducted in the previous section.

5.5.3 Experiments

To empirically validate the well-posedness of perspective uncalibrated photometric stereo under general lighting, we implemented the previous algorithm in MATLAB, and evaluated it against 16 synthetic datasets. These datasets were created by considering four 3D-shapes ("Armadillo", "Bunny", "Joyful Yell", and "Thai Statue"[1]) and four different albedo maps ("White", "Bars", "Ebsd", and "Voronoi"). Ground-truth normals were deduced from the depth maps using (5.12), approximating partial derivatives of the depth with first-order finite differences. Then, for each of the 16 combinations of 3D-shape and albedo, $m = 21$ images were simulated according to (5.20), while varying the lighting coefficient, as illustrated in Fig. 5.2. Each image is of size 1600×1200, and comes along with the ground-truth normals, reconstruc-

---

[1]Joyful Yell: https://www.thingiverse.com/thing:897412; other datasets: http://www-graphics. stanford.edu/data/3dscanrep.

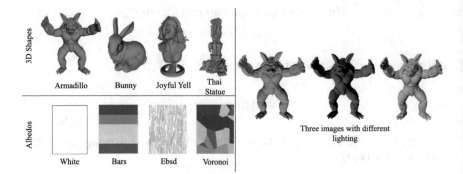

Fig. 5.2 The four 3D-shapes and four albedo maps used to create 16 (3D-shape, albedo) datasets. For each dataset, $m = 21$ images were rendered under varying first-order spherical harmonics lighting. On the right, we show three images of the ("Armadillo", "White") dataset

Table 5.1 Mean angular error (in degrees), for each (3D-shape, albedo) combination. The error remains below 10 degrees for each dataset. This indicates that the genuine geometry is recovered, and empirically confirms the well-posedness of perspective uncalibrated photometric stereo under first-order spherical harmonics lighting

3D-shape	Albedo			
	White	Bars	Ebsd	Voronoi
Armadillo	2.01	1.81	2.13	2.03
Bunny	1.42	1.38	1.63	1.42
Joyful Yell	5.13	5.35	5.46	5.28
Thai Statue	6.33	6.40	6.46	7.62

tion domain Ω, and intrinsic camera parameters (the focal length f, and the principal point used as reference for pixel coordinates). For the evaluation, we measured the mean angular error (in degrees) between the estimated and the ground-truth normals.

As can be seen in Table 5.1, the mean angular error on the estimated normals is very low (less than 10 degrees for all datasets). This confirms that the geometry of the scene is unambiguously estimated. The images being synthesized without any additional noise or outlier to the Lambertian model (e.g., shadows or specularities), one may, however, be surprised that the mean angular error is nonzero. As suggested in [25], the observed residual errors may be due to the finite differences approximation of partial derivatives arising in the perspective integrability matrix (matrix $\mathbf{I}_p$ in (5.61), which contains the partial derivatives of the entries of the initial $\mathbf{m}$-field, cf. (5.51)). In our implementation, we considered first-order finite differences: other choices of finite differences might reduce the error, yet we leave this as a perspective.

Next, we evaluated the robustness of the proposed approach to an increasing amount of zero-mean, Gaussian noise added to the images of the (Armadillo, White) dataset. As can be seen in Table 5.2, the proposed method dramatically fails as soon as the noise becomes really perceptible (here, failure is observed when standard

Table 5.2 Mean angular error (in degrees) on the (Armadillo, White) dataset, with increasing amount of zero-mean Gaussian noise added to the input images. When noise is negligible, the proposed method largely outperforms the state-of-the-art method from [11]. However, it should be discarded in the presence of strong noise

Method	Standard deviation σ (in percents of the maximum intensity)								
	0.00	0.01	0.02	0.04	0.1	0.2	0.3	0.4	0.5
Reference [11]	18.19	18.19	18.19	18.19	18.19	18.19	18.19	18.20	**18.20**
Ours	**2.01**	**2.07**	**2.12**	**2.33**	**2.90**	**4.43**	**6.56**	**9.14**	113.38

deviation $\sigma > 0.5\%$). For comparison, we also provide the results obtained with the state-of-the-art method [11], which is based on heuristical shape initialization followed by regularized nonconvex refinement. The heuristical nature of the initialization induces a non-negligible bias in shape estimation, which is clearly visible on noise-free data. However, this alternative is much more robust to noise.

This is not really surprising, since the proposed method is spectral, and the alternative one is based on evolved nonconvex optimization. In general, the former is faster (in our implementation on a recent computer, our results were obtained in less than 10 s, and the alternative ones in around 30 min.), but the latter is more robust. Similar observations have been made in other computer vision communities, e.g., pose estimation: the 8-point algorithm [19] is usually replaced by bundle adjustment techniques [41] in order to handle the unavoidable noise arising in real-world data. Overall, the proposed algorithm should be considered only as a way to empirically confirm the well-posedness of the problem, yet on real-world data the existing numerical implementations should be preferred.

5.6 Conclusion and Perspectives

We have studied the well-posedness of uncalibrated photometric stereo under general illumination, represented by first-order spherical harmonics. We have established that integrability reduces the scaled Lorentz ambiguity to a global concave/convex ambiguity in the orthographic case, and resolves it in the perspective one. As Table 5.3 summarizes, this generalizes previous results which were restricted to the directional lighting model. Still, open questions remain concerning further generalization of these results to even more evolved lighting models. For instance, future research on the topic may consider the case of unknown second-order spherical harmonics [3], or that of unknown nearby point light sources [26]. Such generalizations would be of interest from a practical perspective, because the former represents natural illumination very accurately [10], and the latter allows using inexpensive light sources such as LEDs [32].

Table 5.3 State-of-the-art of theoretical results concerning the well-posedness of uncalibrated photometric stereo under different lighting models (directional, spherical harmonics of order 1 and 2, or nearby point sources). We indicate the number of degrees of freedoms (dof) of the underlying ambiguity, and how imposing integrability reduces this number under both orthographic and perspective projection. The bold results refer to the findings in the present paper, and the question marks to remaining open problems

Lighting model	Underlying ambiguity	Effect of imposing integrability	
		Orthographic	Perspective
Directional	9-dof (linear) [12]	3-dof (GBR) [44]	Well-posed [25]
SH_1	6-dof (scaled Lorentz) [3]	**1-dof (concave/convex)**	**Well-posed**
SH_2	9-dof (linear) [3]	?	?
Nearby point	4-dof (rotation and scale) [26]	?	?

Appendix 1: Proof of Proposition 5.1

Proposition 5.1 characterizes the integrability of a normal field in terms of the coefficients m_1, m_2 and m_3 of $\mathbf{m} := \rho\mathbf{n}$. The following proof of this proposition is largely inspired by [25].

Proof According to Eqs. (5.15)–(5.17), integrability of the normal field under perspective projection can be written as:

$$\hat{p}_v = \hat{q}_u \text{ over } \Omega,$$

$$\Longleftrightarrow \left(\frac{p}{f-up-vq}\right)_v = \left(\frac{q}{f-up-vq}\right)_u \text{ over } \Omega,$$

$$\Longleftrightarrow fp_v - vqp_v + vpq_v - fq_u + upq_u - uqp_u = 0 \text{ over } \Omega,$$

$$\Longleftrightarrow \begin{pmatrix} p_v \\ q_v \\ 0 \end{pmatrix}^{\mathsf{T}} \begin{pmatrix} 0 \\ -f \\ v \end{pmatrix} \times \begin{pmatrix} p \\ q \\ -1 \end{pmatrix} + \begin{pmatrix} p_u \\ q_u \\ 0 \end{pmatrix}^{\mathsf{T}} \begin{pmatrix} -f \\ 0 \\ u \end{pmatrix} \times \begin{pmatrix} p \\ q \\ -1 \end{pmatrix} = 0 \text{ over } \Omega,$$

$$(5.75)$$

where $\times$ denotes the cross-product.

Besides, $-\dfrac{\mathbf{m}}{m_3} = \begin{pmatrix} p \\ q \\ -1 \end{pmatrix}$ according to (5.14). If we denote $\mathbf{w}_1 = [0, -f, v]^{\mathsf{T}}$ and $\mathbf{w}_2 = [-f, 0, u]^{\mathsf{T}}$, then (5.75) yields the following equation over Ω:

$$\left(\frac{-\mathbf{m}}{m_3}\right)_v^\top \mathbf{w}_1 \times \left(\frac{-\mathbf{m}}{m_3}\right) + \left(\frac{-\mathbf{m}}{m_3}\right)_u^\top \mathbf{w}_2 \times \left(\frac{-\mathbf{m}}{m_3}\right) = 0,$$

$$\Longleftrightarrow \left(-\frac{m_3\mathbf{m}_v - m_{3v}\mathbf{m}}{m_3^2}\right)^\top \mathbf{w}_1 \times \left(\frac{-\mathbf{m}}{m_3}\right) + \left(-\frac{m_3\mathbf{m}_u - m_{3u}\mathbf{m}}{m_3^2}\right)^\top \mathbf{w}_2 \times \left(\frac{-\mathbf{m}}{m_3}\right) = 0. \quad (5.76)$$

Multiplying Eq. (5.76) by m_3^3:

$$(m_3\mathbf{m}_v - m_{3v}\mathbf{m})^\top \mathbf{w}_1 \times \mathbf{m} + (m_3\mathbf{m}_u - m_{3u}\mathbf{m})^\top \mathbf{w}_2 \times \mathbf{m} = 0 \quad \text{over } \Omega. \quad (5.77)$$

In addition, $(\mathbf{w}_1 \times \mathbf{m}) \perp \mathbf{m}$ and $(\mathbf{w}_2 \times \mathbf{m}) \perp \mathbf{m}$, thus the following relationship holds over Ω:

$$m_3\mathbf{m}_v^\top(\mathbf{w}_1 \times \mathbf{m}) + m_3\mathbf{m}_u^\top(\mathbf{w}_2 \times \mathbf{m}) = 0,$$
$$\Longleftrightarrow \mathbf{m}_v^\top(\mathbf{w}_1 \times \mathbf{m}) + \mathbf{m}_u^\top(\mathbf{w}_2 \times \mathbf{m}) = 0,$$
$$\Longleftrightarrow u(m_{1u}m_2 - m_1 m_{2u}) + v(m_{1v}m_2 - m_1 m_{2v})$$
$$+ f(m_{1v}m_3 - m_1 m_{3v}) - f(m_{2u}m_3 - m_2 m_{3u}) = 0. \quad (5.78)$$

which concludes the proof. ☐

Appendix 2: Proof of Theorem 5.1

Theorem 5.1 characterizes scaled Lorentz transformations. Its proof relies on the following Propositions 5.2–5.4 from Lorentz' group theory (proofs of these propositions can be found in [14]).

Proposition 5.2 *For any proper and orthochronous Lorentz transformation* $\mathbf{A} \in L_o^p$, *there exists a unique couple* $(\mathbf{v}, \mathbf{O}) \in B(0, 1) \times SO(3, \mathbb{R})$ *such that*

$$\mathbf{A} = \mathbf{S}(\mathbf{v})\,\mathbf{R}(\mathbf{O}) = \begin{pmatrix} \gamma & \gamma\,\mathbf{v}^\top\mathbf{O} \\ \gamma\,\mathbf{v} & (\mathbf{I}_3 + \frac{\gamma^2}{1+\gamma}\mathbf{v}\mathbf{v}^\top)\mathbf{O} \end{pmatrix}, \quad (5.79)$$

where $\gamma = \frac{1}{\sqrt{1-\|\mathbf{v}\|^2}}$, *and*

$$\mathbf{S}(\mathbf{v}) = \begin{pmatrix} \gamma & \gamma\,\mathbf{v}^\top \\ \gamma\,\mathbf{v} & \mathbf{I}_3 + \frac{\gamma^2}{1+\gamma}\mathbf{v}\mathbf{v}^\top \end{pmatrix}, \quad \mathbf{R}(\mathbf{O}) = \begin{pmatrix} 1\ 0\ 0\ 0 \\ 0 \\ 0 \quad \mathbf{O} \\ 0 \end{pmatrix}. \quad (5.80)$$

Proposition 5.3 *The product of two proper/improper transformations is a proper one, and the product of a proper and improper transformation is an improper one. The same for the orthochronous property.*

Proposition 5.4 *Matrix* $\mathbf{T} = \begin{pmatrix} -1 & 0 & 0 & 0 \\ 0 & 1 & 0 & 0 \\ 0 & 0 & 1 & 0 \\ 0 & 0 & 0 & 1 \end{pmatrix}$ *is improper and non-orthochronous, and*

matrix $\mathbf{P} = \begin{pmatrix} 1 & 0 & 0 & 0 \\ 0 & -1 & 0 & 0 \\ 0 & 0 & -1 & 0 \\ 0 & 0 & 0 & -1 \end{pmatrix}$ *is improper and orthochronous.*

Using these already known results, we propose the following characterization of Lorentz transformations:

Proposition 5.5 *For any Lorentz transformation* $\mathbf{A} \in L$, *there exists a unique couple* $(\mathbf{v}, \mathbf{O}) \in B(\mathbf{0}, 1) \times SO(3, \mathbb{R})$ *such that*

$$\mathbf{A} = \begin{pmatrix} \epsilon_1(\mathbf{A})\, \gamma & \epsilon_1(\mathbf{A})\, \gamma\, \mathbf{v}^\top \mathbf{O} \\ \epsilon_2(\mathbf{A})\, \gamma\, \mathbf{v} & \epsilon_2(\mathbf{A})(\mathbf{I}_3 + \frac{\gamma^2}{1+\gamma}\mathbf{v}\mathbf{v}^\top)\mathbf{O} \end{pmatrix}. \tag{5.81}$$

Proof We first assume that $\mathbf{A} \in L_n^i$.

According to Proposition 5.4, $\mathbf{T} \in L_n^i$. Thus using Proposition 5.3: $\mathbf{TA} \in L_o^p$. Therefore, according to Proposition 5.2, there exists a unique couple $(\mathbf{v}, \mathbf{O}) \in B(\mathbf{0}, 1) \times SO(3, \mathbb{R})$ such that $\mathbf{TA} = \mathbf{S}(\mathbf{v})\mathbf{R}(\mathbf{O})$. Since $\mathbf{TT} = \mathbf{I}_4$, this implies that $\mathbf{A} = \mathbf{TS}(\mathbf{v})\mathbf{R}(\mathbf{O})$. In addition, $\epsilon_1(\mathbf{A}) = -1$ and $\epsilon_2(\mathbf{A}) = 1$, hence:

$$\mathbf{A} = \begin{pmatrix} \epsilon_1(\mathbf{A})\gamma & \epsilon_1(\mathbf{A})\gamma\mathbf{v}^\top\mathbf{O} \\ \epsilon_2(\mathbf{A})\gamma\mathbf{v} & \epsilon_2(\mathbf{A})(\mathbf{I}_3 + \frac{\gamma^2}{1+\gamma}\mathbf{v}\mathbf{v}^\top)\mathbf{O} \end{pmatrix}. \tag{5.82}$$

With the same reasoning, we get the result for all the other transformations. □

Combining Proposition 5.5 and the definition (5.24) of scaled Lorentz transformations, we get Theorem 5.1.

Appendix 3: Some Useful Results on GBR and Lorentz Matrices, and Corollary 5.1

The aim of this section is to prove Corollary 5.1, which was used in the proofs of Theorems 5.2 and 5.3. Its proof relies on a few results on GBR and Lorentz matrices, which we provide in the following.

Let us denote by G the group of GBR transformations, and by G_s that of scaled GBR transformations defined by

$$G_s = \{s\mathbf{A} \mid s \in \mathbb{R}\backslash\{0\} \text{ and } \mathbf{A} \in G\}. \tag{5.83}$$

Both are subgroups of $GL(3, \mathbb{R})$ under the matrix product. For all $\mathbf{B} = s\mathbf{A} \in G_s$, we call s the scale part of $\mathbf{B}$, and $\mathbf{A}$ its GBR part.

Let $\mathbf{C} \in \mathbb{R}^{n\times n}$ with $n > 1$ and C_{ij} its entries. We will use the following notation for a minor of size two:

$$C_{k,l}^{i,j} = C_{ij}C_{kl} - C_{kj}C_{il}, \tag{5.84}$$

where $1 \leq i < k \leq n$ and $1 \leq j < l \leq n$.

Such minors allow to characterize scaled GBR matrices:

Proposition 5.6 *Let $\mathbf{A} \in \mathbb{R}^{3\times3}$, A_{ij} the entries of $\mathbf{A}$. Then, $\mathbf{A}$ is a scaled GBR transformation iff $\mathbf{A}$ is invertible and fulfills the following equations:*

$$\begin{cases} A_{3,2}^{2,1} = A_{3,3}^{2,1} = A_{3,3}^{1,2} = A_{3,2}^{1,1} = 0, \\ A_{3,3}^{2,2} = A_{3,3}^{1,1}. \end{cases} \tag{5.85}$$

Proof See [4]. $\qquad\square$

Proposition 5.7 *Let $\mathbf{v} \in B_0(1)$, $\gamma = \frac{1}{\sqrt{1-\|\mathbf{v}\|^2}}$, then $\mathbf{C} = \mathbf{I_3} + \frac{\gamma^2}{1+\gamma}\mathbf{v}\mathbf{v}^\top$ is positive definite.*

Proof Let $\mathbf{B} = \frac{\gamma^2}{1+\gamma}\mathbf{v}\mathbf{v}^\top$. We note $E_\lambda(\mathbf{B})$ the eigenspace associated to the eigenvalue λ of $\mathbf{B}$. $\mathbf{B}$ is symmetric, thus according to the spectral theorem, all the eigenvalues of $\mathbf{B}$ are real, and $\mathbb{R}^3 = \bigoplus_{i=1}^{r} E_{\lambda_i}(\mathbf{B})$ with $r \leq 3$ the number of eigenvalues, and $\{\lambda_i\}_{i=1...r}$ the eigenvalues of $\mathbf{B}$. Hence: $\dim(\mathbb{R}^3) = \sum_{i=1}^{r} \dim(E_{\lambda_i}(\mathbf{B}))$. According to the rank-nullity theorem, $\dim(Ker(\mathbf{B})) + rank(\mathbf{B}) = 3$, and by definition $rank(\mathbf{B}) = 1$, thus $\dim(Ker(\mathbf{B})) = \dim(E_0(\mathbf{B})) = 2$. We deduce that there exists a unique nonzero eigenvalue $\lambda \in \mathbb{R}\backslash\{0\}$ such that $\mathbb{R}^3 = E_0(\mathbf{B}) \bigoplus E_\lambda(\mathbf{B})$ with $\dim(E_\lambda(\mathbf{B})) = 1$.

Let $\Pi_\mathbf{v}$ the orthogonal projection onto $span\{\mathbf{v}\}$, and let $\mathbf{x} \in \mathbb{R}^3 : \Pi_\mathbf{v}(\mathbf{x}) = \frac{\mathbf{v}\mathbf{v}^\top}{\|\mathbf{v}\|^2}\mathbf{x}$ and $\Pi_\mathbf{v}(\mathbf{v}) = \mathbf{v}$. We have: $\mathbf{B} = \frac{\gamma^2}{1+\gamma}\|\mathbf{v}\|^2 \Pi_\mathbf{v}$ and $\mathbf{B}\mathbf{v} = (\frac{\gamma^2}{1+\gamma}\|\mathbf{v}\|^2)\mathbf{v}$. Thus, $\frac{\gamma^2}{1+\gamma}\|\mathbf{v}\|^2$ is an eigenvalue of $\mathbf{B}$ and $\lambda = \frac{\gamma^2}{1+\gamma}\|\mathbf{v}\|^2$. Besides, $\frac{\lambda}{\gamma-1} = \frac{\gamma^2}{(\gamma-1)(\gamma+1)}\|\mathbf{v}\|^2 = \frac{\gamma^2}{\gamma^2-1}\|\mathbf{v}\|^2 = \frac{1}{1-\frac{1}{\gamma^2}}\|\mathbf{v}\|^2 = \frac{1}{1-(1-\|\mathbf{v}\|^2)}\|\mathbf{v}\|^2 = 1$. Therefore, $\lambda = \gamma - 1$ and the eigenvalues of $\mathbf{B}$ are 0 and $(\gamma - 1)$. Let $\alpha \in \{0, \gamma - 1\}$ and $\mathbf{u} \in E_\alpha(\mathbf{B})$. We have:

$$\mathbf{B}\mathbf{u} = \alpha\mathbf{u} \iff \mathbf{u} + \mathbf{B}\mathbf{u} = \mathbf{u} + \alpha\mathbf{u} \iff \mathbf{C}\mathbf{u} = (\alpha + 1)\mathbf{u}. \tag{5.86}$$

Thus, $1 > 0$ and $\gamma > 0$ are the eigenvalues of $\mathbf{C}$ with $E_1(\mathbf{C}) = E_0(\mathbf{B})$ and $E_\gamma(\mathbf{C}) = E_{\gamma-1}(\mathbf{B})$. Consequently, $\mathbf{C}$ is positive definite. $\qquad\square$

Proposition 5.8 *Let* $\mathbf{A}_s \in L_s$ *a scaled Lorentz transformation. The submatrix* $\mathbf{B}$ *formed by the last 3 rows and 3 columns of* $\mathbf{A}_s$ *is invertible.*

Proof By definition of $\mathbf{A}_s$, there exists a unique couple $(s, \tilde{\mathbf{A}}) \in \mathbb{R}\backslash\{0\} \times L$ such that $\mathbf{A}_s = s\tilde{\mathbf{A}}$. Hence, from Proposition 5.5, there exists a unique couple $(\mathbf{v}, \mathbf{O}) \in B(\mathbf{0}, 1) \times SO(3, \mathbb{R})$ such that $\mathbf{B} = s\epsilon_2(\tilde{\mathbf{A}})(\mathbf{I}_3 + \frac{\gamma^2}{1+\gamma}\mathbf{v}\mathbf{v}^\top)\mathbf{O}$. Since $\mathbf{O} \in SO(3, \mathbb{R})$, $\det(\mathbf{O}) = 1$. In addition, Proposition 5.7 implies $\det(\mathbf{I}_3 + \frac{\gamma^2}{1+\gamma}\mathbf{v}\mathbf{v}^\top) > 0$, thus $\det(\mathbf{B}) \neq 0$. □

Corollary 5.1 *Let* $\mathbf{A}_s \in L_s$ *a scaled Lorentz transformation. If its entries* A_{ij} *fulfill*

$$\begin{cases} A_{4,3}^{3,2} = A_{4,4}^{3,2} = A_{4,4}^{2,3} = A_{4,3}^{2,2} = 0, \\ A_{4,4}^{3,3} = A_{4,4}^{2,2}, \end{cases} \tag{5.87}$$

then the submatrix $\mathbf{B}$ *of* $\mathbf{A}_s$ *formed by the last 3 rows and 3 columns is a scaled GBR, i.e., there exists a unique quadruple* $(\lambda, \mu, \nu, \beta) \in \mathbb{R}^4$ *with* $\lambda \neq 0$, $\beta \neq 0$ *such that*

$$\mathbf{A}_s = \begin{pmatrix} A_{11} & A_{12} & A_{13} & A_{14} \\ A_{21} & \beta\lambda & 0 & -\beta\mu \\ A_{31} & 0 & \beta\lambda & -\beta\nu \\ A_{41} & 0 & 0 & \beta \end{pmatrix}. \tag{5.88}$$

Proof According to Proposition 5.8, $\mathbf{B}$ is invertible. Besides, $\mathbf{A}_s$ fulfill Eqs. (5.87) iff $\mathbf{B}$ fulfill Eqs. (5.85), thus according to Proposition 5.6, $\mathbf{B} \in G_s$. □

References

1. Bartal O, Ofir N, Lipman Y, Basri R (2018) Photometric stereo by hemispherical metric embedding. J Math Imaging Vis 60(2):148–162
2. Basri R, Jacobs DW (2003) Lambertian reflectances and linear subspaces. IEEE Trans Pattern Anal Mach Intell 25(2):218–233
3. Basri R, Jacobs DW, Kemelmacher I (2007) Photometric stereo with general, unknown lighting. Int J Comput Vis 72(3):239–257
4. Belhumeur PN, Kriegman DJ, Yuille AL (1999) The bas-relief ambiguity. Int J Comput Vis 35(1):33–44
5. Breuß M, Cristiani E, Durou J-D, Falcone M, Vogel O (2012) Perspective shape from shading: ambiguity analysis and numerical approximations. SIAM J Imaging Sci 5(1):311–342
6. Chen G, Han K, Shi B, Matsushita Y, Wong K-YK (2019) Self-calibrating deep photometric stereo networks. In: Proceedings of the IEEE conference on computer vision and pattern recognition (CVPR), pp 8739–8747
7. Chen L, Zheng Y, Shi B, Subpa-Asa A, Sato I (2019) A microfacet-based model for photometric stereo with general isotropic reflectance. IEEE Trans Pattern Anal Mach Intell (in press)
8. Durou J-D, Falcone M, Sagona M (2008) Numerical methods for shape-from-shading: a new survey with benchmarks. Comput Vis Image Underst 109(1):22–43
9. Einstein A (1905) Zur elektrodynamik bewegter körper. Annalen der physik 322(10):891–921

10. Frolova D, Simakov D, Basri R (2004) Accuracy of spherical harmonic approximations for images of Lambertian objects under far and near lighting. In: Proceedings of the European conference on computer vision (ECCV), pp 574–587
11. Haefner B, Ye Z, Gao M, Wu T, Quéau Y, Cremers D (2019) Variational uncalibrated photometric stereo under general lighting. In: Proceedings of the international conference on computer vision (ICCV)
12. Hayakawa H (1994) Photometric stereo under a light source with arbitrary motion. J Opt Soc Am A 11(11):3079–3089
13. Ikehata S (2018) CNN-PS: CNN-based photometric stereo for general non-convex surfaces. In: Proceedings of the European conference on computer vision (ECCV), pp 3–18
14. Jaffe A (2013) Lorentz transformations, rotations, and boosts, 2013. Online notes. http://home.ku.edu.tr/~amostafazadeh/phys517_518/phys517_2016f/Handouts/A_Jaffi_Lorentz_Group.pdf. Accessed Sept 2019
15. Khanian M, Boroujerdi AS, Breuß M (2018) Photometric stereo for strong specular highlights. Comput Visual Media 4(1):83–102
16. Kozera R, Prokopenya A (2018) Second-order algebraic surfaces and two image photometric stereo. In: International conference on computer vision and graphics, pp 234–247
17. Li J, Robles-Kelly A, You S, Matsushita Y (2019) Learning to minify photometric stereo. In: Proceedings of the IEEE conference on computer vision and pattern recognition (CVPR), pp 7568–7576
18. Logothetis F, Mecca R, Quéau Y, Cipolla R (2016) Near-field photometric stereo in ambient light. In: Proceedings of the British machine vision conference (BMVC)
19. Longuet-Higgins HC (1987) A computer algorithm for reconstructing a scene from two projections. In: Fischler MA, Firschein O (eds) Readings in computer vision. Morgan Kaufmann, Burlington, pp 61–62
20. Mecca R, Falcone M (2013) Uniqueness and approximation of a photometric shape-from-shading model. SIAM J Imaging Sci 6(1):616–659
21. Mecca R, Wetzler A, Bruckstein AM, Kimmel R (2014) Near field photometric stereo with point light sources. SIAM J Imaging Sci 7(4):2732–2770
22. Mecca R, Quéau Y, Logothetis F, Cipolla R (2016) A single-lobe photometric stereo approach for heterogeneous material. SIAM J Imaging Sci 9(4):1858–1888
23. Mo Z, Shi B, Lu F, Yeung S-K, Matsushita Y (2018) Uncalibrated photometric stereo under natural illumination. In: Proceedings of the IEEE conference on computer vision and pattern recognition (CVPR), pp 2936–2945
24. Onn R, Bruckstein A (1990) Integrability disambiguates surface recovery in two-image photometric stereo. Int J Comput Vis 5(1):105–113
25. Papadhimitri T, Favaro P (2013) A new perspective on uncalibrated photometric stereo. In: Proceedings of the IEEE conference on computer vision and pattern recognition (CVPR), pp 1474–1481
26. Papadhimitri T, Favaro P (2014) Uncalibrated near-light photometric stereo. In: Proceedings of the British machine vision conference (BMVC)
27. Peng S, Haefner B, Quéau Y, Cremers D (2017) Depth super-resolution meets uncalibrated photometric stereo. In: Proceedings of the IEEE international conference on computer vision (ICCV) workshops, pp 2961–2968
28. Poincaré H (1905) Sur la dynamique de l'électron. Comptes Rendus de l'Académie des Sciences 140:1504–1508
29. Quéau Y, Mecca R, Durou J-D, Descombes X (2017) Photometric stereo with only two images: a theoretical study and numerical resolution. Image Vis Comput 57:175–191
30. Quéau Y, Wu T, Cremers D (2017) Semi-calibrated near-light photometric stereo. In: International conference on scale space and variational methods in computer vision (SSVM), pp 656–668
31. Quéau Y, Wu T, Lauze F, Durou J-D, Cremers D (2017) A non-convex variational approach to photometric stereo under inaccurate lighting. In: Proceedings of the IEEE conference on computer vision and pattern recognition (CVPR), pp 99–108

32. Quéau Y, Durix B, Wu T, Cremers D, Lauze F, Durou J-D (2018) LED-based photometric stereo: modeling, calibration and numerical solution. J Math Imaging Vis 60(3):313–340
33. Quéau Y, Durou J-D, Aujol J-F (2018) Normal integration: a survey. J Math Imaging Vis 60(4):576–593
34. Radow G, Hoeltgen L, Quéau Y, Breuß M (2019) Optimisation of classic photometric stereo by non-convex variational minimisation. J Math Imaging Vis 61(1):84–105
35. Ramamoorthi R, Hanrahan P (2001) An efficient representation for irradiance environment maps. In: Proceedings of the annual conference on computer graphics and interactive techniques, pp 497–500
36. Sengupta S, Zhou H, Forkel W, Basri R, Goldstein T, Jacobs D (2018) Solving uncalibrated photometric stereo using fewer images by jointly optimizing low-rank matrix completion and integrability. J Math Imaging Vis 60(4):563–575
37. Shi B, Inose K, Matsushita Y, Tan P, Yeung S-K, Ikeuchi K (2014) Photometric stereo using internet images. In: Proceedings of the international conference on 3D vision (3DV), vol 1, pp 361–368
38. Shi B, Wu Z, Mo Z, Duan D, Yeung S-K, Tan P (2019) A benchmark dataset and evaluation for non-Lambertian and uncalibrated photometric stereo. IEEE Trans Pattern Anal Mach Intell 41(2):271–284
39. Tozza S, Falcone M (2016) Analysis and approximation of some shape-from-shading models for non-Lambertian surfaces. J Math Imaging Vis 55(2):153–178
40. Tozza S, Mecca R, Duocastella M, Del Bue A (2016) Direct differential photometric stereo shape recovery of diffuse and specular surfaces. J Math Imaging Vis 56(1):57–76
41. Triggs B, McLauchlan PF, Hartley RI, Fitzgibbon AW (1999) Bundle adjustment – a modern synthesis. In: International workshop on vision algorithms (ICCV workshops), pp 298–372
42. Woodham RJ (1978) Reflectance map techniques for analyzing surface defects in metal castings. Technical report AITR-457. MIT
43. Woodham RJ (1980) Photometric method for determining surface orientation from multiple images. Opt Eng 19(1):139–144
44. Yuille A, Snow D (1997) Shape and albedo from multiple images using integrability. In: Proceedings of the IEEE conference on computer vision and pattern recognition (CVPR), pp 158–164
45. Zhang R, Tsai P-S, Cryer JE, Shah M (1999) Shape-from-shading: a survey. IEEE Trans Pattern Anal Mach Intell 21(8):690–706

Chapter 6
Recent Progress in Shape from Polarization

Boxin Shi, Jinfa Yang, Jinwei Chen, Ruihua Zhang, and Rui Chen

Abstract Photometric cues play an important role in recovering per-pixel 3D information from images. Shape from shading and photometric stereo are popular photometric 3D reconstruction approaches, which rely on inversely analyzing an image formation model of the surface normal, reflectance, and lighting. Similarly, shape from polarization explores radiance variation under different polarizer angles to estimate the surface normal, which does not require an active light source and has less restricted assumptions on reflectance and lighting. This chapter reviews basic principles of shape from polarization and its image formation model for surfaces of different reflection properties. We then survey recent progress in shape from polarization combined with different auxiliary information such as geometric cues, spectral cues, photometric cues, and deep learning, and further introduce how polarization imaging benefits other vision tasks in addition to shape recovery.

B. Shi · J. Chen
National Engineering Laboratory for Video Technology, Department of Computer
Science and Technology, Peking University, Beijing 100871, China
e-mail: shiboxin@pku.edu.cn

J. Chen
e-mail: cjw@pku.edu.cn

J. Yang (✉)
School of Electronics Engineering and Computer Science, Department of Machine Intelligence,
Peking University, Beijing 100871, China
e-mail: jinfayang@pku.edu.cn

R. Zhang · R. Chen
School of Earth and Space Science, Institute of Remote Sensing and GIS,
Peking University, Beijing 100871, China
e-mail: ruihuazhang@pku.edu.cn

R. Chen
e-mail: rui.chen@pku.edu.cn

© Springer Nature Switzerland AG 2020
J.-D. Durou et al. (eds.), *Advances in Photometric 3D-Reconstruction*,
Advances in Computer Vision and Pattern Recognition,
https://doi.org/10.1007/978-3-030-51866-0_6

6.1 Introduction

3D sensing and reconstruction is an important research topic in computer vision. Depending on the key constraint where 3D information is obtained, 3D reconstruction approaches can be divided into two categories according to the shape estimation cues they exploit. Generally speaking, the geometric approach (e.g., multi-view stereo (MVS) [21, 22]) extracts corresponding points across images from different viewpoints to reconstruct the 3D surface using triangulation; the reconstructed surface may not contain accurate 3D information when the correspondence is unreliably detected in regions with few textures. In contrast, the photometric approach (e.g., shape from shading (SfS) [26], photometric stereo (PS) [35, 58]) infers the per-pixel surface normal by analyzing the radiance variation in different lighting conditions; the recovered surface normal map is dense and contains rich and detailed geometric information even if in textureless regions, but further integration is required to reconstruct the surface. This chapter will mainly discuss photometric approaches.

Analysis of how the light interacts with surface geometry is the key to photometric 3D reconstruction. The classic SfS or PS inversely resolves the rendering equation with a shading term (dot product of surface normal and lighting) scaled by a reflectance term (Lambertian albedo) using single or multiple images under calibrated distant light sources. Recent research on SfS or PS focuses on how to generalize reflectance and/or lighting assumptions to extend their practicability on real scenarios. According to a recent survey on photometric stereo [58], around one hundred images illuminated by different distant light sources with calibrated intensity and direction are required for a reliable estimation of the surface normal of objects with general isotropic reflectance.

The polarization information of the reflected light also contains useful information about the surface normal of an object. The method of shape recovery by polarization information (Shape-from-Polarization, SfP) was first proposed by Shurcliff in 1962 [12, 59]. Intuitively speaking, the unpolarized light reflected from a surface point becomes partially polarized, and the observed scene radiance varies with changing the polarizer angle, which encodes some relationship with the surface normal. Therefore, by analyzing such relationship at each surface point, SfP recovers per-pixel surface normal with as high resolution as a modern 2D image sensor, which is similar to SfS and PS. Compared to SfS or PS, SfP has the following advantages [30, 31]:

- Suitable for materials with different reflection properties: Polarization information is widely observed in a broad category of materials, including dielectrics, metals, and even transparent objects.
- Weak assumption on light sources: As long as the incident light is unpolarized and becomes partially polarized after reflecting over the surface, useful cues for 3D reconstruction would be observed.
- Passive acquisition: A typical SfP approach captures the radiance variation under different polarizations by rotating a polarizer in front of the lens under natural light.

The first and second advantages above complement the major challenges about strict reflection and lighting assumptions when applying SfS or PS. The third advantage shows the great potential of designing a high-resolution 3D camera without relying on active light sources (like active stereo or structured light). Despite the advantages listed above, the mapping from radiance variation to surface normal is not one to one. Hence, to obtain the correct surface normal we have to solve the ambiguity [43] by exploring additional constraints from various aspects. Since surface normal is the only output for most of the SfP approaches, auxiliary information from geometric prior is particularly helpful in guiding the surface from gradient recovery which is also a challenge for SfS and PS. In this chapter, we first review the principles of polarization images (Sect. 6.2) and how classic SfP works for surfaces with different types of reflection properties (Sect. 6.3). We then discuss the recent progress in SfP based on the additional information from four different sources (Sect. 6.4):

- SfP + geometric cues: The auxiliary geometric information can be used to reduce the ambiguity of SfP and guide the surface normal integration. Such information can be obtained from MVS [5, 17, 43, 44, 46, 72] or depth sensors [30, 31, 75].
- SfP + spectral cues: Cameras with multiple spectral bands provide additional information of the scene, which can be utilized by SfP for disambiguation and estimating the refractive index of the surface [27, 28, 45, 62, 70, 71].
- SfP + photometric cues: The normal estimates from SfS [7, 49, 67] or PS [4, 8] have different ambiguities or no ambiguity, which can complement SfP normal estimates, or directly estimate surface height by solving a linear system [60, 61, 67, 73].
- SfP + deep learning: Comprehensive priors learned from a data-driven approach using deep neural networks also enables SfP [9] to deal with real-world scenarios, which deviate from ideal physics models.

We further introduce various applications of polarization imaging to other vision tasks in addition to 3D shape recoveries, such as image segmentation, robot navigation, image enhancement, and reflection separation (Sect. 6.5) before concluding this chapter (Sect. 6.6) with discussions on open problems.

6.2 Principles of Polarization Imaging

This chapter begins with a brief introduction of the principles of polarization imaging, with a special focus on image formation models used for SfP. The symbols and their corresponding physical meanings used in this chapter are summarized in Table 6.1.

Table 6.1 Symbols and notations used in this chapter

Symbol	Physical meaning
λ	Wavelength
$\mathbf{u}$	Pixel location $\mathbf{u}(x, y)$
$R_\perp$ and $R_\parallel$	Fresnel reflection coefficient
$T_\perp$ and $T_\parallel$	Fresnel refraction coefficient
S_0, S_1, S_2	Stokes component
ϑ_{pol}	Polarizer angle
η_i and η_t	Refractive index of the original medium and the refracted medium
$\mathbf{n}(\mathbf{u})$	Normal vector $\mathbf{n}$ at pixel $\mathbf{u}$
ρ and ϕ	degree of polarization and polarization phase angle
$\alpha(\mathbf{u})$ and $\theta(\mathbf{u})$	Azimuth and zenith angles of surface normal at pixel $\mathbf{u}$

6.2.1 Principles of Fresnel Reflection

Fresnel theory describes the radiance of incident light on the surface, reflected light, and refracted light. It is the theoretical basis for studying SfP and other vision problems using polarization imaging. For most commercial cameras, the recorded scene radiance is nonlinearly mapped to the pixel value according to the radiometric response function. Throughout this chapter, we assume the camera has a linear radiometric response function (or it is radiometrically calibrated), and use intensity for brevity.

The light wave is a transverse wave, and vibration direction of the transverse wave is perpendicular to the propagation direction. As shown in Fig. 6.1, when a light wave travels from medium 1 to a uniform smooth medium 2, some of the light penetrates the surface and gets refracted. At the same time, affected by the light wave electric field inside the medium, the electrons in the reflecting medium at the surface will vibrate, form a dipole, and radiate outward, generating reflected light. The intensity of the reflected light depends on the intensity of the electron vibration perpendicular to the direction along with the reflected light travels.

The plane where the incident light, the refracted light, and the reflected light are defined as the incident plane. Light vectors that vibrate in any direction can be decomposed into two mutually perpendicular components [36, 59]. The energy radiance of the incident light wave can be decomposed into two parts, a component perpendicular to the incident plane E_{is} and a component parallel to the incident plane E_{ip}. Similarly, reflected light and refracted light can be decomposed into E_{rs}, E_{rp}, E_{ts}, and E_{tp}, as shown in Fig. 6.1. For parallel components, the direction of vibration of the induced dipole is generally not orthogonal to the reflected light, directly leading to energy attenuation. The dipole vibration direction of the vertical component is always orthogonal to the reflected light, so there is no attenuation. The two orthogonal components of the reflected light E_{rs} and E_{rp} have different values of amplitude. For these reasons, when the incident light is unpolarized, the reflected

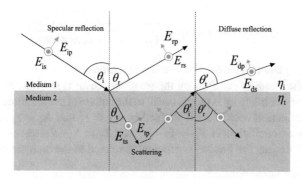

Fig. 6.1 Illustration of Fresnel reflection (rearranged based on Fig. 2 in [7]). From medium 1 to medium 2, θ_i, θ_r, and θ_t represent angle of incidence, angle of reflection, and angle of refraction, respectively. Similarly, θ_i', θ_r', and θ_t' are the corresponding angles from medium 2 to medium 1. Please refer to Table 6.1 for the definitions of notations

light will become partially polarized, i.e., the superposition of the unpolarized part and the completely polarized part.

The following equation holds when light is refracted in two media with refractive indices of η_i and η_t, respectively:

$$\eta_i \sin \theta_i = \eta_t \sin \theta_t. \tag{6.1}$$

This equation is called the refraction law of light, also known as Snell's Law.

At the interface of the medium, the components of the electric field and the magnetic field in the tangential direction are continuously distributed, so that the amplitudes of the incident light and the reflected light in medium 1 and the amplitude of the refracted light in medium 2 are equal in the same direction. According to this principle, the reflection process shown in Fig. 6.1, i.e., the amplitude of reflection coefficient perpendicular to the direction of the incident plane, can be defined as

$$r_s(\eta_i, \eta_t, \mu_i, \mu_t, \theta_i) \equiv \frac{E_{rs}}{E_{is}} = \frac{\frac{\eta_i}{\mu_i} \cos \theta_i - \frac{\eta_t}{\mu_t} \cos \theta_t}{\frac{\eta_i}{\mu_i} \cos \theta_i + \frac{\eta_t}{\mu_t} \cos \theta_t}, \tag{6.2}$$

where μ_i and μ_t are the magnetic permeability of the two mediums, respectively. To simplify the problem, we can assume $\mu_i = \mu_t = \mu_0$ (μ_0 is vacuum permeability). θ_i is the angle of incidence, and θ_t is the angle of refraction. Therefore, Eq. (6.2) can be simplified into

$$r_s(\eta_i, \eta_t, \mu_i, \mu_t, \theta_i) \equiv \frac{E_{rs}}{E_{is}} = \frac{\eta_i \cos \theta_i - \eta_t \cos \theta_t}{\eta_i \cos \theta_i + \eta_t \cos \theta_t}. \tag{6.3}$$

Similarly, the amplitude of reflection coefficient parallel to the incident plane can be expressed as follows:

$$r_\mathrm{p}(\eta_\mathrm{i}, \eta_\mathrm{t}, \mu_\mathrm{i}, \mu_\mathrm{t}, \theta_\mathrm{i}) \equiv \frac{E_\mathrm{rp}}{E_\mathrm{ip}} = \frac{\eta_\mathrm{t} \cos \theta_\mathrm{i} - \eta_\mathrm{i} \cos \theta_\mathrm{t}}{\eta_\mathrm{t} \cos \theta_\mathrm{i} + \eta_\mathrm{i} \cos \theta_\mathrm{t}}. \tag{6.4}$$

Equations (6.3) and (6.4) are called Fresnel equations.

Usually, the intensity received by the imaging sensor is not the amplitude of the light wave, and the intensity is proportional to the square of the amplitude. Therefore, the Fresnel reflection coefficient is defined as $R_\perp = r_\mathrm{s}^2$ and $R_\parallel = r_\mathrm{p}^2$. Accordingly, the transmission coefficient is $T_\perp = 1 - R_\perp$ and $T_\parallel = 1 - R_\parallel$, as can be seen from the above discussion, the Fresnel coefficient is related to the incident angle of light.

6.2.2 Polarization Image Formation Model

The previous section discusses the Fresnel principle on which the polarization application is based. We will now explain how to apply the above theory to polarization images. After the unpolarized light is reflected by the surface of an object, there are two types of reflection: specular reflection and diffuse reflection. The polarization information can be defined by using three physical quantities: the light intensity I, polarization phase angle ϕ, and degree of polarization ρ, which represents the proportion of polarized light in the reflected light (Wolff [69] considered partially polarized light as the sum of fully polarized and unpolarized light).

The polarization information of the incident light reflected by the object can be captured by using a polarization filter (a.k.a. polarizer). A linear polarizer can be placed in front of the camera and rotated to different angles to obtain polarization images at different angles. When rotating the polarizer, the intensity of each pixel on the acquired image will also change, as shown in Fig. 6.2.

Fig. 6.2 Transmitted radiance sinusoid (TRS). For two pixels (A and B) with different surface normals, the observed intensities under varying polarizer angles form different sinusoid curves

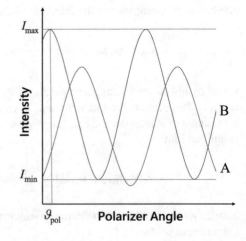

Specifically, the intensity of each pixel will show a sinusoidal variation with rotation angle of the polarizer, which is called Transmitted Radiance Sinusoid (TRS) curve [52, 65]. If we denote the change of brightness value with polarizer angle as $I_{\vartheta_{\text{pol}}}$, TRS can be expressed as

$$I_{\vartheta_{\text{pol}}} = \frac{I_{\max} + I_{\min}}{2} + \frac{I_{\max} - I_{\min}}{2} \cos(2\vartheta_{\text{pol}} - 2\phi), \tag{6.5}$$

where $I_{\max}$ and $I_{\min}$ represent the maximum and minimum brightness values observed during the rotation of the polarizer, respectively. The reference direction of the rotation of the polarizer is defined on a situational basis. For example, when the rotation angle is the same as the polarization phase angle, the maximum brightness is obtained, $I(\vartheta_{\text{pol}} = \phi) = I_{\max}$; then the minimum value is taken by rotating the polarizer by 90° w.r.t. the maximum angle, $I(\vartheta_{\text{pol}} = \phi \pm 90°) = I_{\min}$. Since the cosine function has a period of 2π, Eq. (6.5) always holds for different phase angles ϕ with a difference of 180°. This is also called π-ambiguity as mentioned in [17].

The degree of polarization (DoP) is defined as

$$\rho = \frac{I_{\max} - I_{\min}}{I_{\max} + I_{\min}}. \tag{6.6}$$

By taking a sequence of polarization images, the parameters in TRS can be determined by solving optimization problems [6, 27, 69]. For example, Wolff et al. [69] used three images I_0, I_{45}, and I_{90}, which are captured by rotating the polarizer to 0°, 45°, and 90°, respectively. Atkinson et al. [4] introduced to use four polarization images I_0, I_{45}, I_{90}, and I_{135} to estimate the Stokes parameters, expressed as a 4-vector $S = [S_0, S_1, S_2, S_3]^T$, where S_0 is the power of the incident beam, S_1 and S_2 represent the power of 0° and 45° linear polarization, and S_3 is the power of right circular polarization. For unpolarized natural light, the value of S_3 is small and is often ignored. The Stokes parameters can be obtained by rotating the polarizer to multiple polarization angles to obtain a set of images:

$$\begin{aligned} S_0 &= \frac{I_0 + I_{45} + I_{90} + I_{135}}{2} \\ S_1 &= I_0 - I_{90} \\ S_2 &= I_{45} - I_{135}, \end{aligned} \tag{6.7}$$

and then fit the solution of the polarization parameters as

$$\begin{aligned} I &= S_0 \\ \phi &= \frac{1}{2} \arctan_2(S_2, S_1) \\ \rho &= \frac{\sqrt{S_1^2 + S_2^2}}{S_0}, \end{aligned} \tag{6.8}$$

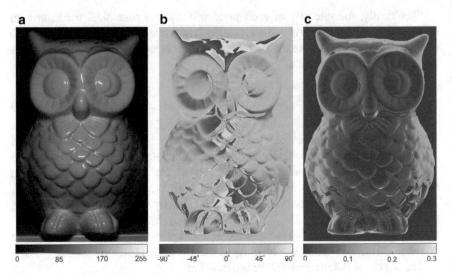

Fig. 6.3 **a** One of the input images. **b** Polarization phase angle. **c** Degree of polarization

where arctan$_2$ is the four-quadrant arctangent operator [14].

An example of visualization for polarization parameters obtained from a sequence of polarization images is shown in Fig. 6.3. The target object is a white ceramic owl.

6.2.3 Polarization Data Acquisition

The most straightforward way to capture polarization images is to add a rotatable linear polarizer in front of the lens of an ordinary camera. The illustration of such a capturing system can be found in Fig. 6.4a. The approaches using a rotating filter are "division of time" and they can use the full resolution of the sensor but trade-off against acquisition time.

Wolff et al. [69] used liquid crystal polarization cameras instead of ordinary cameras to capture a complete set of polarization data, including information such as the degree of polarization at each pixel location at the frame rate of about one frame per second, which makes it possible to apply polarization capture for moving targets. More recently, it becomes possible to capture a set of polarization images using a single shot [1, 19, 39]. These cameras have an array of micro-polarizers in front of the CCD or using multiple CCDs behind several polarizers with a shared lens to simultaneously capture images under different polarizer angles, as shown in Fig. 6.4b. Other polarization cameras follow a similar way of data acquisition [50, 51], and these cameras usually provide corresponding data display and analysis software to visualize and process the captured polarization data (such as DoP) for various applications. Embedding micro-polarizers in front of the imaging sensor is

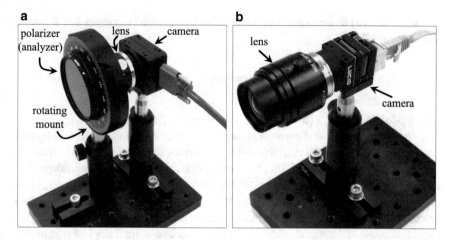

Fig. 6.4 Two typical ways of capturing polarization images: **a** an ordinary camera (Point Gray CM3-U3-50S5C-CS) with a linear polarizer being put in front of the lens ("division of time"). **b** The polarization camera (Lucid Vision PHX050S-PC) with micro-polarizers in front of the imaging sensor ("division of focal plane")

"division of focal plane" that allows instantaneous capture, but such an approach sacrifices the spatial resolution of the imaging sensor.

6.3   Shape from Polarization for Surfaces of Different Reflection Properties

After polarization information (the intensity, polarization phase angle, degree of polarization, etc.) is captured by using the devices and methods mentioned above, the goal of SfP is to estimate the surface normal (azimuth angle and zenith angle) by analyzing the relationship between the polarization image formation model and normal. The coordinate system used in polarization imaging usually aligns the Z-axis direction with the reflected light entering the camera direction. The normal vector can be represented by its azimuth and zenith angle in a spherical coordinate system. The azimuth angle α ($0° \leq \alpha \leq 360°$) of normal is defined as the angle between the projection of the incident plane on the X-Y imaging plane and the X-axis. The zenith angle θ ($0° \leq \theta \leq 90°$) of normal is defined as the angle between the observation and the surface normal direction. Given the azimuth and zenith angles, the normal vector at any point on the surface can be expressed as

$$\mathbf{n}(\mathbf{u}) = [n_x(\mathbf{u}), n_y(\mathbf{u}), n_z(\mathbf{u})]^\top$$
$$= [\sin \alpha(\mathbf{u}) \sin \theta(\mathbf{u}), \cos \alpha(\mathbf{u}) \sin \theta(\mathbf{u}), \cos \theta(\mathbf{u})]^\top. \tag{6.9}$$

Denoting $z(\mathbf{u})$ to represent a surface, Eq. (6.9) can also be written in the form of surface gradient as

$$\mathbf{n}(\mathbf{u}) = \frac{1}{\sqrt{p(\mathbf{u})^2 + q(\mathbf{u})^2 + 1}}[-p(\mathbf{u})^2 \ -q(\mathbf{u})^2 \ 1]^\top, \qquad (6.10)$$

where $p(\mathbf{u}) = \partial z_x(\mathbf{u})$, $q(\mathbf{u}) = \partial z_y(\mathbf{u})$, i.e., $\nabla z(\mathbf{u}) = [p(\mathbf{u}) \ q(\mathbf{u})]^\top$.

After obtaining the surface normal map, the surface can be reconstructed by integrating the surface normal map represented as a two-dimensional second-order continuous differential function [20], i.e., combination of partial derivatives independent of order of differential variables, $\frac{\partial^2 f(x,y)}{\partial x \partial y} = \frac{\partial^2 f(x,y)}{\partial y \partial x}$. There are many sophisticated approaches [2, 3, 20, 34, 48] for solving the "gradient to surface" problem, however, a detailed survey of integrating normal to surface is beyond the scope of this chapter. Note that most SfP methods only produce the surface normal map as output. But when a coarse geometry is provided like "SfP + Geometric Cues" [30, 31] introduced later or a partial differential formulation is applied [37, 60, 61, 73], the output becomes height map.

6.3.1 Shape from Specular Polarization

The specular reflection is the mirror-like reflection behavior caused by the body reflection on top of the surface, as shown in Fig. 6.1. It provides useful information about the surface geometry as indicated by 3D reconstruction approaches analyzing the mirror-like reflection behavior without considering the polarization property [29]. If the polarization property of reflected light is taken into consideration, the relationship between the Fresnel coefficient and specular angle of incidence can be plotted as shown in Fig. 6.5. When the reflected light is polarized in the direction parallel to the incident plane, its attenuation will be greater, which can be denoted as $R_\perp > R_\parallel$. Therefore, when the polarizer is placed perpendicular to the azimuth angle of the surface, the strongest reflected light can be obtained (we assume that the image is captured under orthographic projection). At this point, the azimuth angle of normal α could take the value of either $\phi + \frac{\pi}{2}$ or $\phi + \frac{3\pi}{2}$.

The value of zenith angle of normal can be inferred by looking at the degree of polarization. According to Eq. (6.6), the degree of polarization can be directly determined by I_{max} and I_{min}. According to Fig. 6.5a,

$$I_{max} = \frac{R_\perp}{R_\perp + R_\parallel} I_s, \quad I_{min} = \frac{R_\parallel}{R_\perp + R_\parallel} I_s, \qquad (6.11)$$

where I_s is the specular reflection intensity (assuming there is no diffuse reflection). Combining Eq. (6.6) with Eq. (6.11) gives the degree of specular polarization expressed by the Fresnel coefficient:

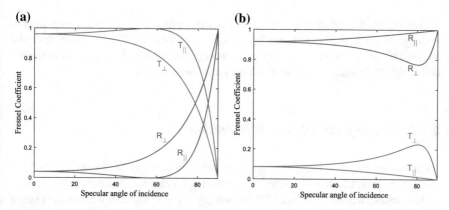

Fig. 6.5 The relationship between Fresnel coefficient and angle of incidence for dielectric object (left) and metallic object (right, $\eta = 0.8, k = 6$) [12]. Note the difference of Fresnel coefficients between diffuse (related to T) and specular reflection (related to R) results in different formula of degree of polarization

$$\rho_s = \frac{R_\perp(\eta, \theta_i) - R_\|(\eta, \theta_i)}{R_\perp(\eta, \theta_i) + R_\|(\eta, \theta_i)}. \tag{6.12}$$

By further combining Eq. (6.3) with Eq. (6.4), the degree of polarization expressed by the refractive index η and the zenith angle θ are finally obtained as

$$\rho_s = \frac{2 \sin^2 \theta \cos \theta \sqrt{\eta^2 - \sin^2 \theta}}{\eta^2 - \sin^2 \theta - \eta^2 \sin^2 \theta + 2 \sin^4 \theta}. \tag{6.13}$$

So far, the zenith angle of normal for specular surface can be calculated by Eq. (6.13) with an analytical solution provided in [30].

The specular reflection theory for polarization can be further extended to the surface reconstruction of metallic objects [47]. However, the metal properties are different from those of the dielectric, as shown in Fig. 6.5b, mainly because the refractive index of metal is more complicated. The metal refractive index $\hat{\eta}$ can be defined as $\hat{\eta} = \eta(1 + i\kappa)$, where κ is the attenuation coefficient. To reduce computational complexity, the following approximation is usually made

$$|\hat{\eta}|^2 = \eta^2(1 + \kappa^2) \gg 1. \tag{6.14}$$

Similarly, according to the approximation of the metal refractive index, the relationship between the degree of polarization of the metal surface and the zenith angle is as

$$\rho(\theta) = \frac{2\eta \cdot \tan \theta \sin \theta}{\tan^2 \theta \sin^2 \theta + |\hat{\eta}|^2}. \tag{6.15}$$

Related studies of specular reflection can be extended to the study of transparent objects, such as [44, 53]; besides, studies on specular reflection roughness and anisotropy can be found in [23, 24]; more recent research based on specular reflection shows the application of SfP to black specular surface [46].

6.3.2 Shape from Diffuse Polarization

As shown in Fig. 6.1, except for specular reflection on top of the surface body, a portion of the light enters the interior of the object, gets refracted during the process and becomes partially polarized [70]. According to Fig. 6.5a, the polarized light being parallel to the incident plane has the maximum transmission coefficient, which can be denoted as $T_\parallel > T_\perp$. Therefore, when the polarizer is placed parallel to the incident plane, the maximum reflected light intensity can be observed. This is exactly the opposite case of the specular reflection for dielectric surface (see Fig. 6.5a). Finally, the azimuth angle of normal α can take the value of either ϕ or $\phi \pm \pi$.

The scattering of light inside the object is irregular, so the light coming out of the object becomes polarized light with intensity attenuation. Similar to the process of light entering the medium from the air, when the light goes back to the air from the medium (see Fig. 6.1), its relative refractive index becomes $1/\eta$ instead of η (refractive index for air is 1). The refraction angle of the refracted ray corresponding to the incident light at any angle can be solved by Snell's law. Further, the Fresnel transmission coefficient for a given transmission angle can be obtained from Fig. 6.5a.

It should be noted that the calculation of the degree of polarization for diffuse reflection is based on the transmission coefficient ($T_\parallel$ and $T_\perp$) (light transmitted from interior of the object medium then to the air medium), rather than the reflection coefficient ($R_\parallel$ and $R_\perp$) (light from the air medium refraction then to the air medium). According to Eq. (6.6), we have

$$\rho_d = \frac{T_\parallel(1/\eta, \theta_i') - T_\perp(1/\eta, \theta')}{T_\parallel(1/\eta, \theta_i') + T_\perp(1/\eta, \theta')}$$
$$= \frac{R_\perp(1/\eta, \theta_i') - R_\parallel(1/\eta, \theta_i')}{2 - R_\perp(1/\eta, \theta_i') - R_\parallel(1/\eta, \theta_i')}. \tag{6.16}$$

Substituting Eqs. (6.3) and (6.4) into Eq. (6.16), the DoP of diffuse surface becomes

$$\rho_d = \frac{(\eta - 1/\eta)^2 \sin^2 \theta}{2 + 2\eta^2 - (\eta + 1/\eta)^2 \sin^2 \theta + 4 \cos \theta \sqrt{\eta^2 - \sin^2 \theta}}. \tag{6.17}$$

From the equation above, the zenith angle of normal can be calculated in an analytic form [6].

6.3.3   Shape from Mixed Polarization

The study of specular and diffuse polarization above considers their characteristics separately. Compared with specular reflection, diffuse reflection has a lower signal-to-noise ratio thus is more difficult to measure useful information. Besides, according to Fig. 6.6, the influence of the refractive index of specular reflection is less than that of diffuse reflection. However, diffuse reflection also has its unique advantages. For example, the relationship between the degree of polarization and zenith angle is monotonic for diffuse reflection, which is not the same case for specular reflection, as shown in Fig. 6.6. Since specular polarized reflection is usually observed for material with strong specular reflectance property, when the surface normal bisects the viewing and lighting directions. It is not easy to find a dielectric object that only contains strong specular reflectance, while a diffuse reflectance usually exists and is more widely observed.

In a real-world scenario, the surface reflectance usually contains mixture reflectance of diffuse and specular components. The dichromatic reflection model can be adapted to modeling such mixture reflectance by assuming a linear superposition of specular and diffuse reflections [25, 41, 66], i.e., "Diffuse + Specular" $(\mathcal{D} + \mathcal{S})$. SfP approach usually assumes that either diffuse or specular component is dominant and then uses the conclusion in Sects. 6.3.1 and 6.3.2 to solve the problem [30]. By further assuming that for a pixel $\mathbf{u}$ it either belongs to the specular reflection point set $\mathcal{S}$, or belongs to the diffuse reflection point set $\mathcal{D}$ (the complete point set is $\mathcal{F} = \mathcal{D} \cup \mathcal{S}$), Taamazyan et al. [63] proposed an SfP solution for mixed polarization. The specular information and the polarization information can be simultaneously separated from the specular and diffuse reflection, and at the same time, the refractive index of the

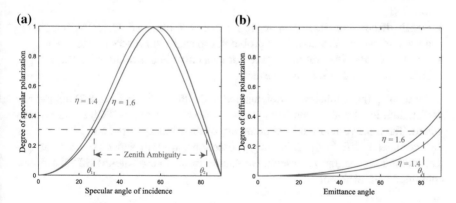

Fig. 6.6 Relationship between the specular angle of incidence/emittance angle and degree of polarization. The left figure shows the specular case: the same DoP corresponds to two angles of incidence and influence of the index of refraction is not obvious. The right figure shows the diffuse case: DoP corresponds to emittance angle in a one-to-one mapping and influence of the index of refraction is stronger than that of specular case

object can be recovered. Cui [17] derived an interesting proposition: Under the mixed polarization, no matter which kind of polarized light dominates, the light intensity information measured at any point on the surface is always sinusoidal, i.e., Eq. (6.5) is independent of the proportion of specular or diffuse reflection in the reflected light. Based on this proposition, Cui et al. [17] obtained a high-precision 3D surface for mixed polarization surfaces, assisted by rough geometry estimated from multi-view images.

6.4   Shape from Polarization + "X"

According to the analysis in Sects. 6.2 and 6.3, we can tell that by only using images and polarization information, it is difficult to obtain unique solution for surface normal estimates. For zenith angle, if there is specular component in surface reflectance, its solution will be not unique (see Fig. 6.6). For azimuth angle, its estimation contains different ambiguities for diffuse and specular surfaces w.r.t. polarization phase angle ϕ: (1) For diffuse reflection, the relationship is ϕ or $\phi + \pi$; (2) for specular reflection, the relationship is $\phi + \frac{\pi}{2}$ or $\left(\phi + \frac{\pi}{2}\right) + \pi$, i.e., there is a $\frac{\pi}{2}$ offset from the diffuse case, which is called $\frac{\pi}{2}$-ambiguity in [16, 17].

In addition to the inherent $\pi/2$ and π azimuthal ambiguity, SfP using only polarization images also suffer from other limitations as summarized in [6, 30, 31]:

- **Refractive distortion**: Obtaining zenith component of the surface normal requires knowledge of the refractive index of the object, which is usually unknown.
- **Fronto-parallel surfaces**: When the zenith angle is close to zero, the degree of polarization is small and the estimated normals are noisy in these regions due to low SNR.
- **Depth discontinuities**: Even if the normals are obtained correctly, integrating normal to surface is a nontrivial problem, especially for discontinuous depth.
- **Relative depth**: Integrating surface normals only obtains relative 3D shape, up to unknown offset and scaling constants.

To resolve these inherent problems and make the solution from SfP unique and useful, additional constraints from various aspects (denoted as "X") could be integrated. Table 6.2 provides a brief summary of their input, output and how different "X" completes SfP. In the following subsections, we will show examples of typical methodologies and representative results from state-of-the-art SfP algorithms combined with different "X".

Table 6.2 Comparison of different "SfP + X" methods

Method	Input	Output	How to complement SfP
SfP + Geometric Cues [5, 17, 30, 31, 43, 44, 46, 72, 75]	Pol. images from multiple viewpoints	Normal + depth	Shape ambiguity, normal integration
SfP + Spectral Cues [27, 28, 45, 62, 70, 71]	Pol. images from a multi-spectral camera	Normal	Shape ambiguity, Refractive index
SfP + Photometric Cues [4, 7, 8, 49, 60, 61, 67, 73]	Pol. images under varying lightings	Normal (+ depth [60, 61])	Shape ambiguity
SfP + Deep Learning [9]	Pol. images	Normal	Shape ambiguity

6.4.1 Shape from Polarization + Geometric Cues

Ambiguities in SfP can be corrected if an initially obtained geometric information is provided (this geometry could be rough), which could be obtained by multi-view stereo methods or low-cost depth sensors.

Multiview stereo: Atkinson et al. [5] used a binocular stereo setup to get polarization images and geometric cues. Miyazaki et al. [43, 44, 46] took polarization images from multiple (more than three) viewpoints and combine specular SfP to model transparent surfaces. Cui et al. [17] combined SfP with MVS. The ambiguous azimuth angle from SfP was used to improve MVS for textureless regions during an iso-depth contour tracing process which naturally bypassed the π-ambiguity. Yang et al. [72] further extended [17] by capturing data using a single-shot polarization camera and replacing MVS with monocular SLAM, so that real-time 3D reconstruction using polarization became possible. A more recent combination of SfP and three-view geometry is introduced in [16].

Depth sensor: Kadambi et al. [30, 31] used a Kinect camera to obtain a rough depth map for removing the ambiguity in azimuth angle, and guided the normal to surface reconstruction operation. Zhu et al. [75] used an RGB camera to estimate the albedo map and a stereo camera to obtain geometry cues.

We take [30, 31] as an example for further explanation, whose pipeline is shown in Fig. 6.7.

The Kinect sensor is first used to obtain a rough depth map $\mathbf{D} \in \mathbb{R}^{M \times N}$. Note the initial depth is severely quantified and contains much noise. Such an initial depth map is then converted to surface normal domain as $\mathbf{N}^{\text{depth}} \in \mathbb{R}^{M \times N \times 3}$. A baseline SfP method is applied to images taken with different polarizer angles, which contains the π-ambiguity in azimuth angle (assuming the surface reflectance is diffuse dominant). The normal calculated from the initial depth map can be used to correct the ambiguity. Defining a binary linear operator $\mathcal{A} \in \{0, 1\}$, the relationship between the two normal

(a) Input: Kinect Only
Microsoft Kinect Version II

(b) Input: Polarization Photos
Canon T3i DSLR
Hoya CIR-PL Filter

(c) Shape from Polarization

(d) Coarse depth to correct
azimuthal ambiguity artifacts

(e) Correcting refractive distortion
and physics-based integration

Fig. 6.7 Pipeline of the "SfP + Geometric Cues" method in [30]. **a** The rough depth of an object is captured by Kinect and **b** three photos are captured by rotating a polarizer in front of the camera. **c** Integration of surface normals obtained from Fresnel equations (baseline SfP). Note the π-ambiguity (observed as a flip in the shape) and distortion of the zenith angle (observed as flat regions in the middle of the cup). **d** Integration of surface normals after correcting the π-ambiguity removes the flip (note the details in shape brought by high-resolution surface normal estimates from SfP). The final result is shown in **e** after correcting for zenith distortion and guiding the integration using the rough depth in **a**

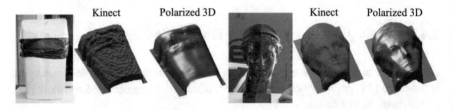

Kinect   Polarized 3D   Kinect   Polarized 3D

Fig. 6.8 Example result of the "SfP + Geometric Cues" method in [30]

maps can be formulated as the following optimization problem:

$$\widehat{\mathcal{A}} = \arg\min_{\mathcal{A}} \left\| \mathbf{N}^{\text{depth}} - \mathcal{A}\left(\mathbf{N}^{\text{polar}}\right) \right\|_2^2 + \gamma \|\nabla \mathcal{A}\|_1, \tag{6.18}$$

where $\|\nabla \mathcal{A}\|_1$ is the smoothing constraint added to the total variation problem, γ controls the smoothing effect of the solution pixel by pixel. Then the optimal operator $\widehat{\mathcal{A}}$ is obtained by using the graph cut method, and is applied to the normal from baseline SfP $\mathbf{N}^{\text{corr}} = \widehat{\mathcal{A}}\left(\mathbf{N}^{\text{polar}}\right)$ for disambiguation. Given the optimized surface normal, the surface $\widehat{\mathbf{D}} \in \mathbb{R}^{M \times N}$ is reconstructed by using a spanning tree constrained integration, which fuses the normal (high-frequency pixel-wise details) and depth (low-frequency geometric positions).

An example result from this method is shown in Fig. 6.8. The left example is taken under the unknown environment lighting, while the right example is a specular object containing spatially varying specular reflectance. Such lighting and reflectance bring difficulty in applying SfS or PS, but SfP guided by a rough depth successfully estimates 3D information with rich details.

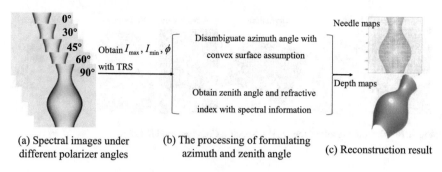

(a) Spectral images under different polarizer angles

(b) The processing of formulating azimuth and zenith angle

(c) Reconstruction result

Fig. 6.9 Pipeline of the "SfP + Spectral Cues" method in [28] (rearranged)

6.4.2 Shape from Polarization + Spectral Cues

The spectral information captured by a multi-band camera contains useful cues related to material properties of the object. Such information can not only assist in disambiguation for SfP, but also help to estimate the refractive index of the surface (note that to use Eqs. (6.13), (6.15), and (6.17), the refractive index of the object needs to be known). The pipeline of [28] can be seen in Fig. 6.9.

Wolff et al. [70, 71] analyzed different kinds of rough and smooth dielectric surfaces and provided some speculation on how to combine these diffuse reflectance models. Miyazaki et al. [45] used visible and far-infrared polarization information for obtaining surface orientations of transparent surfaces. The degree of polarization in the far-infrared wavelength was used for resolving the ambiguity in the visible wavelength. Huynh et al. [27, 28] proposed a method for estimating the zenith angle and surface refractive index directly from the spectral variation of the polarization information. Stolz et al. [62] solved ambiguity in zenith angle with wavelengths of 472 and 655 nm.

To further understand how "SfP + Spectral Cues" works, we take the method proposed in [27, 28] as an example to explain in detail. In order to constrain the change of the refractive index in the wavelength domain, the Cauchy dispersion Eq. [13] was used. The relationship between the refractive index and the spectral band is described by

$$\eta(\mathbf{u}, \lambda) = \sum_{k=1}^{M} C_k(\mathbf{u}) \lambda^{-2(k-1)}, \tag{6.19}$$

where $\mathbf{u}$ represents each pixel in the image, λ represents the wavelength, and M represents the total number of coefficients. It can be seen from Eq. (6.19) that the refractive index η is only related to the band wavelength λ and the dispersion equation coefficient $C_k(\mathbf{u})$. The refractive index can be expressed as a polynomial function of the band. Then the problem of solving the refractive index becomes solving the coefficient of the Cauchy dispersion equation.

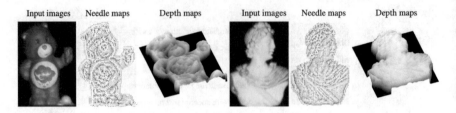

Input images   Needle maps   Depth maps   Input images   Needle maps   Depth maps

Fig. 6.10 Example result of the "SfP + Spectral Cues" method in [28]

The maximum and minimum intensity values at each pixel of the polarization image with the band wavelength λ can be obtained, and with the Fresnel Equation (6.2) we can obtain the following equation:

$$\frac{I_{\min}}{I_{\max}} = \frac{1 - R_\perp(\mathbf{u}, \lambda)}{1 - R_\parallel(\mathbf{u}, \lambda)} = \left[\frac{\cos\theta_{\mathrm{d}}(\mathbf{u})\sqrt{\eta(\mathbf{u}, \lambda)^2 - \sin^2\theta_{\mathrm{d}}(\mathbf{u})} + \sin^2\theta_{\mathrm{d}}(\mathbf{u})}{\eta(\mathbf{u}, \lambda)} \right]^2,$$
(6.20)

where $\theta_{\mathrm{d}}(\mathbf{u})$ represents the zenith angle of normal at pixel $\mathbf{u}$; $\eta(\mathbf{u}, \lambda)$ represents the refractive index at the corresponding pixel point; $R_\perp(\mathbf{u}, \lambda)$ represents the vertical reflection coefficient in the Fresnel equation; $R_\parallel(\mathbf{u}, \lambda)$ represents the parallel reflection coefficient in the Fresnel equation.

Solving Eq. (6.20) can be formulated as a least-squares problem, whose loss function is defined as

$$E(\mathbf{u}) = \sum_{i=1}^{N} \left[\frac{\cos\theta_{\mathrm{d}}(\mathbf{u})\sqrt{\eta(\mathbf{u}, \lambda)^2 - \sin^2\theta_{\mathrm{d}}(\mathbf{u})} + \sin^2\theta_{\mathrm{d}}(\mathbf{u})}{\eta(\mathbf{u}, \lambda)} - r(\mathbf{u}, \lambda_i) \right]^2,$$
(6.21)

where N is the number of bands. The Cauchy dispersion equation is added as the refractive index constraint in this equation. With this constraint, the above loss function becomes an equation with $M + 1$ variables, and the unknowns to be solved are $\theta_{\mathrm{d}}(\mathbf{u})$ and $C_k(\mathbf{u}), k = 1, \ldots, M$. When the number of coefficients of the dispersion equation M satisfies $M + 1 \leq N$, the above nonlinear least-squares problem can be solved by linear search or confidence interval method. After the zenith angle is obtained, shape of the object is obtained by integrating the estimated surface normal. The refractive index η can also be obtained by using the coefficient $C_k(\mathbf{u})$ of Cauchy dispersion equation.

An example result from this method is shown in Fig. 6.10. Images of Bear and Statue are the 45° polarization images. Needle maps reveal the normal orientation along shading contours, and note that the depth maps and the refractive index of objects are simultaneously recovered by this method.

6.4.3   Shape from Polarization + Photometric Cues

Photometric 3D reconstruction approaches such as SfS and PS also estimate the surface normal. The former is a highly ill-posed problem which contains much ambiguity in the estimated surface normal, while the latter also provides ambiguous solutions when the lighting condition is uncalibrated. Fortunately, the ambiguity spaces of SfS, PS, and SfP are not completely overlapped, thus can complement each other in obtaining unique surface normal estimates.

Atkinson et al. [4, 7, 8] used three different known light positions for photometric stereo, and disambiguated the normals given by SfP. Ngo et al. [49] took shading as a constraint to recover the shape of a smooth dielectric object. The shading constraint came from two light directions at the same polarizer angle and polarization constraint came from two different polarizer angles of the same light direction. Tozza et al. [67] described the shading information and polarization information by using partial differential equations and solved the linear differential problem to get object surface height without computing surface normals independently. Yu et al. [73] minimized the sum of squared residuals between predicted and observed intensities over all pixels and polarizer angles, which was solved by nonlinear least-squares optimization. The object albedo and lighting condition were required for initialization. Smith et al. [60, 61] used a large, sparse system of linear equations to solve for surface height directly. The unpolarized intensity provided shading information and served as an additional constraint on the surface normal direction via an appropriate reflectance model.

We take the method in [60, 61] as an example to expand its details, whose pipeline is shown in Fig. 6.11. They assume a diffuse surface followed the Lambertian model with uniform albedo. Then the unpolarized intensity is related to the surface normal by

$$\mathbf{u} \in \mathcal{D} \Rightarrow i_{\mathrm{un}}(\mathbf{u}) = \cos(\theta_{\mathrm{i}}(\mathbf{u})) = \mathbf{n}(\mathbf{u}) \cdot \mathbf{s}, \tag{6.22}$$

where $\theta_{\mathrm{i}}(\mathbf{u})$ is the angle between lighting and normal directions, $\mathbf{s}$ is the light source vector. In terms of the surface gradient, this equation can be written as

$$i_{\mathrm{un}}(\mathbf{u}) = \frac{-p(\mathbf{u})s_x - q(\mathbf{u})s_y + s_z}{\sqrt{p(\mathbf{u})^2 + q(\mathbf{u})^2 + 1}}, \tag{6.23}$$

This method formulates the polarization constraints as linear equations. For a diffuse object, the collinearity condition is used as the constraint for azimuth angle $f(\rho(\mathbf{u}), \eta)$. This condition satisfies either of the two possible azimuth angles implied by the polarization phase angle measurement. Then the linear equation is obtained in the surface gradient with nonlinear normalization factor canceled as

$$\frac{i_{\mathrm{un}}(\mathbf{u})}{f(\rho(\mathbf{u}), \eta)} = -p(\mathbf{u})s_x - q(\mathbf{u})s_y + s_z. \tag{6.24}$$

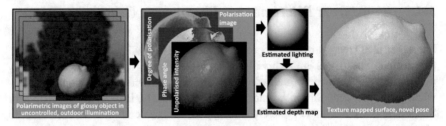

Fig. 6.11 Pipeline of the "SfP + Photometric Cues" method in [61]. From a single polarization image under unknown illumination, the lighting information is estimated and the surface height is directly calculated by combining SfP and the photometric image formation model

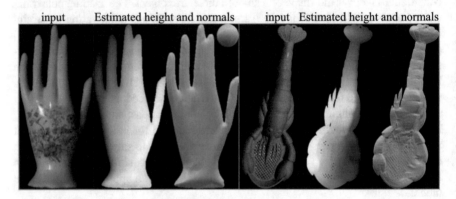

Fig. 6.12 Example result of the "SfP + Photometric Cues" method in [61]

In order to combine the shading constraints and polarization constraints, a least-squares solution over all observed pixels is used, and the optimized results with minimal residual is chosen as the solution. A global optimization is used to find results from random initialization.

An example result from this method is shown in Fig. 6.12. The left shows the estimated height and normals given a texture-mapped object with varying albedo, while the right is the result for an indoors object with point light source and uniform albedo. Note that all results come directly from a single polarization image after solving the linear system.

6.4.4 Shape from Polarization + Deep Learning

Many physics-based vision problems like SfS and PS also get great improvement when combining with deep learning [15]. It will be quite interesting to see how SfP benefits from combining with deep learning. Most of the existing SfP solutions purely rely on the physics-based image formation model (under certain assumptions),

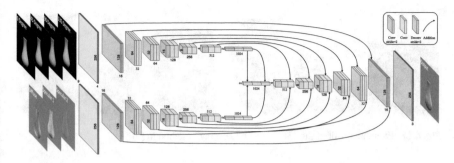

Fig. 6.13 Pipeline of the "SfP + Deep Learning" method in [9]. Features from polarization images and ambiguous surface normals are mutually interacted and refined in an encoder–decoder processing, which are all in a fully convolutional manner

the performance drop becomes significant if real-world conditions deviate from the ideally physics-based model. At present, there are few published results in this field. We mainly introduce the research work of Ba et al. [9], which is the first trial to use deep learning techniques for solving SfP problem as shown in Fig. 6.13

For training a deep neural network for solving SfP, the datasets are crucial. Ba et al.'s method in [9] is also the first attempt to build a real-world dataset training and testing deep SfP networks. They use a polarization camera to capture the polarization images (with polarizer angles in $\{0°, 45°, 90°, 135°\}$) and a laser scanner to obtain the "ground truth" normal being aligned to the image coordinate system (similar to the PS dataset in [58]). Such data are fed into a physics-inspired neural network for regressing the surface normal as

$$\hat{N} = f\left(I_{\phi_1}, I_{\phi_2}, \ldots, I_{\phi_M}, N_{\text{diff}}, N_{\text{spec1}}, N_{\text{spec2}}\right), \qquad (6.25)$$

where $f(\cdot)$ is the prediction model, $\left\{I_{\phi_1}, I_{\phi_2}, \ldots, I_{\phi_M}\right\}$ is a set of polarization images, and $\hat{N}$ is the estimated surface normals. N_{diff} and N_{spec} are surface normals calculated by applying diffuse and specular SfP solutions, respectively. Then a tensor with dimensionality defined as $4 \times H \times W$ is used to describe the polarization images (note the training data captured by a polarization camera uses 4 images as a set), where $H \times W$ is the spatial resolution of images. The loss function for surface normal regression is defined as

$$L_{\text{cosine}} = \frac{1}{W \times H} \sum_{i}^{W} \sum_{J}^{H} \left(1 - \left\langle \hat{N}_{\mathbf{u}}, N_{\mathbf{u}} \right\rangle\right), \qquad (6.26)$$

where $\langle \cdot, \cdot \rangle$ denotes the dot product, $\hat{N}_{\mathbf{u}}$ is the estimated surface normal at pixel location $\mathbf{u}$ (i, j), and $N_{\mathbf{u}}$ is the corresponding ground truth of surface normal. The loss is minimized when $\hat{N}_{\mathbf{u}}$ and $N_{\mathbf{u}}$ have identical orientation. The network is designed according to the encoder–decoder architecture in a fully convolutional manner. Due

Fig. 6.14 Example result of the "SfP + Deep Learning" method in [9]

to the difficulty of capturing large-scale real-world training data, the proposed method does not completely rely on learning from polarization images (upper branch), instead, it also learned from physics-based solutions from non-learning methods (lower branch) to assist the data-driven solution for achieving reliable estimation on complicated real scenarios.

An example result from this method is shown in Fig. 6.14. The physics-inspired deep learning method shows great performance on a shiny object with strong specular reflectance. But as the authors commented, this method still failed on an object with mixed polarization. This could be an interesting open research topic on how to combine SfP and deep learning in a more comprehensive manner.

6.5 Other Applications of Polarization Imaging

The RGB images from an ordinary camera cannot capture the polarization information, i.e., the different distributions for the direction of light fluctuation is not recorded. Analyzing polarization images allows us to explore additional information in light transport, therefore it supports various applications that are not feasible by using only RGB images. In addition to SfP + "X" for high-resolution and precision 3D reconstruction, several applications using other aspects of polarization images will be briefly discussed below.

Image segmentation. It is necessary to use the dividing line of the sky and the ground as a reference to achieve dynamic positioning of a UAV during flight. Traditional methods use color information or edge detection and extraction methods, while Shabayek et al. [57] proposed to use natural polarized light information for this task. This method is less influenced by the very strong sunlight in images thanks to the availability of polarization information, thus achieves more robust segmentation.

Robot navigation. Polarization information can also be combined with binocular stereo for depth estimation of a highly specular scene, which can be applied to robot dynamic navigation [10, 11]. A single-shot polarization camera is required to achieve real-time 3D sensing for robot motion planning and navigation.

Image enhancement. Underwater images usually contain complex reflection, which reduces the visibility of the target object. Schechner et al. [54, 55] proposed the concept of P^4 (Polarization-picture post-processing) to analyze the inverse process of underwater physical image formation model, which relies on different polarization components.

Reflection separation. The reflections caused by common semi-reflectors, such as glass windows, not only affect the image quality but also downgrade the performance of machine vision algorithms. The polarizer is a commonly adopted filter to perform reflection and transmission layer separation if the semi-reflector is a planer and the polarizer is rotated to an angle that maximally suppresses the reflections (a.k.a. Brewster angle). Lyu et al. [40] proposed a deep learning approach to separate these two layers with a pair of unpolarized and polarized images, which simplifies the previous approach that requires three images from different polarizer angles [33, 56, 68].

6.6   Discussion on Open Problems

The research on SfP is still an active area today. For example, in the year of 2019 when this chapter has been finished from submission through revision, there are many papers using polarization imaging being published, to name a few [9, 18, 37, 38, 40, 42, 64, 75], spanning from mathematical analysis to image-based SfP [37] to novel "SfP + X" solution with a pair of stereo image [75]. We conclude this chapter by pointing out some open problems and summarizing several aspects from which future SfP solutions could probably be benefited.

Physics. The complicated light transport property with polarization being considered is one of the biggest challenges in developing purely physics-based SfP solution. Although interesting ratio-invariant property about mixed polarization has been analyzed [17], it is still difficult to analytically derive a parametric image formation model for SfP with mixed polarization. A unified forward representation that explores the explicit relationship between diffuse and specular polarizations is desired for developing physics-based inverse problems of surface normal estimation from polarization.

Learning. Deep learning for SfP is still a less explored area. Part of the reason is that a ray tracer supports complete polarization properties is not publicly available and capturing sufficiently large-scale real data is challenging as well. The first trial in [9] shows the great potential of data-driven SfP solution in filling in the blank of physics. However, it also heavily relies on physics-based solution as initialization. We hope future learning-based SfP could take a better trade-off between physics (as guidance to ensure correctness of physics) and learning (as prior to increase robustness in the wild), not only for SfP from polarization images, but also to improve "SfP + X" methods by integrating successful experiences in learning-based photometric stereo [15, 74] or multi-view stereo solutions [32].

Fusion. The "SfP + X" analysis in Sect. 6.4 of this chapter reveals that the SfP mutually benefits other 3D reconstruction such as shape from shading/photometric stereo [4, 7, 8, 60, 61, 73] and SfM [17, 72]. Table 6.2 also shows that different 3D reconstruction approaches complement SfP from different aspects. Thus, a truly all-powerful 3D reconstruction approach for arbitrary scenes does not exist. To achieve more accurate and reliable 3D reconstruction, future works should pay attention to fusion-based solutions that combine photometric and geometric, active and passive, depth and normal, as well as physics and learning, to achieve joint improvement.

Sensor. The popularity of polarization cameras in recent years also provide a great possibility for putting SfP into practice. Without a polarization camera for efficient data capture, it is impossible to achieve polarization-based SLAM in real time [72] or capturing plenty of real data for training a deep neural network for SfP [9]. The camera manufactures are also actively working on narrowing the gap between polarization images and ordinary RGB images, e.g., by enabling color capture in polarization imaging with a single shot [39]. We believe in the near future there could be novel types of polarization camera with higher resolution, better SNR, or more different polarizer angles. The emergence of such sensors will definitely push forward the research on SfP in terms of both efficiency and accuracy.

Like other physics-based 3D modeling approaches such as photometric stereo, SfP produces less reliable accuracy if properties of real object reflectance and environment lighting do not fit the theoretical model well. Therefore, it is still not straightforward to apply conventional SfP for 3D reconstruction in the wild, such as directly applying it to a mobile phone. However, the unique properties of polarization, such as the DoP is of high sensitivity to transparent surface [43–45], make it ready to be deployed on visual part inspection or product quality assurance system for special materials.

Despite the physical restrictions inherent in SfP (e.g., unpolarized light source assumption, difficulties in handling mixture polarization, and so on) and the shape estimation ambiguity caused, the development of "SfP + X" and learning-based solution significantly reduces the ambiguity and relaxes the reliance on strict physics assumptions. With more researchers draw their attention to this area, we believe SfP in the wild will finally be realized in the future, probably in a form (but not restricted to it) like the camera on a mobile phone that is able to capture both near and far scene for its high-resolution 3D information in a single shot.

References

1. 4DTechnology: Polarcam snapshot micropolarizer cameras. https://www.4dtechnology.com/products/imaging-polarimeters/
2. Agrawal A, Chellappa R, Raskar R (2005) An algebraic approach to surface reconstruction from gradient fields. In: Proceedings of international conference on computer vision, vol 1. IEEE, pp 174–181
3. Agrawal A, Raskar R, Chellappa R (2006) What is the range of surface reconstructions from a gradient field? In: Proceedings of European conference on computer vision, pp 578–591
4. Atkinson GA (2017) Two-source surface reconstruction using polarisation. In: Scandinavian conference on image analysis, pp 123–135
5. Atkinson GA, Hancock ER (2005) Multi-view surface reconstruction using polarization. In: Proceedings of international conference on computer vision, vol 1. IEEE, pp 309–316

6. Atkinson GA, Hancock ER (2006) Recovery of surface orientation from diffuse polarization. IEEE Trans Image Process 15(6):1653–1664
7. Atkinson GA, Hancock ER (2007) Shape estimation using polarization and shading from two views. IEEE Trans Pattern Anal Mach Intell 29(11):2001–2017
8. Atkinson GA, Hancock ER (2007) Surface reconstruction using polarization and photometric stereo. In: International conference on computer analysis of images and patterns, pp 466–473
9. Ba Y, Chen R, Wang Y, Yan L, Shi B, Kadambi A (2019) Physics-based neural networks for shape from polarization. arXiv preprint arXiv:1903.10210
10. Berger K, Voorhies R, Matthies L (2016) Incorporating polarization in stereo vision-based 3D perception of non-Lambertian scenes. In: Unmanned systems technology, vol 9837. International Society for Optics and Photonics
11. Berger K, Voorhies R, Matthies LH (2017) Depth from stereo polarization in specular scenes for urban robotics. In: IEEE international conference on robotics and automation. IEEE, pp 1966–1973
12. Born M, Wolf E (1997) Principles of optics. Sixth (corrected) edition
13. Born M, Wolf E, Bhatia AB, Clemmow PC, Gabor D, Stokes AR, Taylor AM, Wayman PA, Wilcock WL (1999) Principles of optics: electromagnetic theory of propagation, interference and diffraction of light, 7th edn. Cambridge University Press, Cambridge
14. Burger W, Burge MJ (2009) Principles of digital image processing: fundamental techniques. Springer Publishing Company, Incorporated
15. Chen G, Han K, Shi B, Matsushita Y, Wong KYK (2019) SDPS-net: self-calibrating deep photometric stereo networks. In: Proceedings of computer vision and pattern recognition
16. Chen L, Zheng Y, Subpa-Asa A, Sato I (2018) Polarimetric three-view geometry. In: Proceedings of European conference on computer vision, pp 20–36
17. Cui Z, Gu J, Shi B, Tan P, Kautz J (2017) Polarimetric multi-view stereo. In: Proceedings of computer vision and pattern recognition, pp 1558–1567
18. Cui Z, Larsson V, Pollefeys M (2019) Polarimetric relative pose estimation. In: Proceedings of international conference on computer vision, pp 2671–2680
19. FluxData: Fluxdata polarizatoin camera. http://www.fluxdata.com/products/fd-1665p-imaging-polarimeter
20. Frankot RT, Chellappa R (1988) A method for enforcing integrability in shape from shading algorithms. IEEE Trans Pattern Anal Mach Intell 10(4):439–451
21. Fuhrmann S, Langguth F, Goesele M (2014) MVE-A multi-view reconstruction environment. In: Eurographics workshops on graphics and cultural heritage, pp 11–18
22. Furukawa Y, Ponce J (2009) Accurate, dense, and robust multiview stereopsis. IEEE Trans Pattern Anal Mach Intell 32(8):1362–1376
23. Ghosh A, Chen T, Peers P, Wilson CA, Debevec P (2009) Estimating specular roughness and anisotropy from second order spherical gradient illumination. Comput Graph Forum 28:1161–1170
24. Ghosh A, Chen T, Peers P, Wilson CA, Debevec P (2010) Circularly polarized spherical illumination reflectometry. ACM Trans Graph (TOG) 29(6):162
25. Herrera SEM, Malti A, Morel O, Bartoli A (2013) Shape-from-polarization in laparoscopy. In: IEEE international symposium on biomedical imaging, pp 1412–1415
26. Horn BK (1970) Shape from shading: a method for obtaining the shape of a smooth opaque object from one view
27. Huynh CP, Robles-Kelly A, Hancock E (2010) Shape and refractive index recovery from single-view polarisation images. In: Proceedings of computer vision and pattern recognition, pp 1229–1236
28. Huynh CP, Robles-Kelly A, Hancock ER (2013) Shape and refractive index from single-view spectro-polarimetric images. Int J Comput Vis 101(1):64–94
29. Ihrke I, Kutulakos KN, Lensch HP, Magnor M, Heidrich W (2010) Transparent and specular object reconstruction. Comput Graph Forum 29:2400–2426
30. Kadambi A, Taamazyan V, Shi B, Raskar R (2015) Polarized 3D: high-quality depth sensing with polarization cues. In: Proceedings of international conference on computer vision, pp 3370–3378

31. Kadambi A, Taamazyan V, Shi B, Raskar R (2017) Depth sensing using geometrically constrained polarization normals. Int J Comput Vis 125(1–3):34–51
32. Kar A, Häne C, Malik J (2017) Learning a multi-view stereo machine. In: Advances in neural information processing systems, pp 365–376
33. Kong N, Tai YW, Shin JS (2013) A physically-based approach to reflection separation: from physical modeling to constrained optimization. IEEE Trans Pattern Anal Mach Intell 36(2):209–221
34. Kovesi P (2005) Shapelets correlated with surface normals produce surfaces. In: Proceedings of international conference on computer vision, vol 2. IEEE, pp 994–1001
35. Li M, Zhou Z, Wu Z, Shi B, Diao C, Tan P (2020) Multi-view photometric stereo: a robust solution and benchmark dataset for spatially varying isotropic materials. IEEE Trans Image Process
36. Liao Y (2003) Polarization optics. China Science Press
37. Logothetis F, Mecca R, Sgallari F, Cipolla R (2019) A differential approach to shape from polarisation: a level-set characterisation. Int J Comput Vis 127(11–12):1680–1693
38. Lu J, Ji Y, Yu J, Ye J (2019) Mirror surface reconstruction using polarization field. In: Proceedings of international conference on computational photography. IEEE, pp 1–9
39. LucidVisionLabs: Lucid polarization camera. https://thinklucid.com/polarized-camera-resource-center/
40. Lyu Y, Cui Z, Li S, Pollefeys M, Shi B (2019) Reflection separation using a pair of unpolarized and polarized images. In: Annual conference on neural information processing systems
41. Mecca R, Logothetis F, Cipolla R (2018) A differential approach to shape from polarization. British machine vision conference
42. Miyazaki D, Furuhashi R, Hiura S (2020) Shape estimation of concave specular object from multiview polarization. J Electron Imag 29(4):041006
43. Miyazaki D, Kagesawa M, Ikeuchi K (2003) Polarization-based transparent surface modeling from two views. In: Proceedings of international conference on computer vision, vol 3, p 1381
44. Miyazaki D, Kagesawa M, Ikeuchi K (2004) Transparent surface modeling from a pair of polarization images. IEEE Trans Pattern Anal Mach Intell 26(1):73–82
45. Miyazaki D, Saito M, Sato Y, Ikeuchi K (2002) Determining surface orientations of transparent objects based on polarization degrees in visible and infrared wavelengths. JOSA A 19(4):687–694
46. Miyazaki D, Shigetomi T, Baba M, Furukawa R, Hiura S, Asada N (2016) Surface normal estimation of black specular objects from multiview polarization images. Opt Eng 56(4):041303
47. Morel O, Meriaudeau F, Stolz C, Gorria P (2005) Polarization imaging applied to 3D reconstruction of specular metallic surfaces. Mach Vis Appl Ind Inspect XIII 5679:178–186
48. Nehab D, Rusinkiewicz S, Davis J, Ramamoorthi R (2005) Efficiently combining positions and normals for precise 3D geometry. ACM Trans Graph (TOG) 24(3):536–543
49. Ngo TT, Nagahara H, Taniguchi RI (2015) Shape and light directions from shading and polarization. In: Proceedings of computer vision and pattern recognition, pp 2310–2318
50. PhotonicLattice: Photonic lattice polarizatoin camera. https://www.photonic-lattice.com/en/products/polarization_camera/pi-110/
51. RicohPolar, Ricoh polarizatoin camera. https://www.ricoh.com/technology/tech/051_polarization.html
52. Riviere J, Reshetouski I, Filipi L, Ghosh A (2017) Polarization imaging reflectometry in the wild. ACM Trans Graph (TOG) 36(6):206
53. Saito M, Sato Y, Ikeuchi K, Kashiwagi H (2001) Measurement of surface orientations of transparent objects using polarization in highlight. Syst Comput Jpn 32(5):64–71
54. Schechner YY (2011) Inversion by P^4: polarization-picture post-processing. Philos Trans R Soc B: Biol Sci 366(1565):638–648
55. Schechner YY (2015) Self-calibrating imaging polarimetry. In: Proceedings of international conference on computational photography. IEEE, pp 1–10
56. Schechner YY, Shamir J, Kiryati N (2000) Polarization and statistical analysis of scenes containing a semireflector. JOSA A 17(2):276–284

57. Shabayek AER, Demonceaux C, Morel O, Fofi D (2012) Vision based UAV attitude estimation: progress and insights. J Intell Robot Syst 65(1–4):295–308
58. Shi B, Mo Z, Wu Z, Duan D, Yeung SK, Tan P (2019) A benchmark dataset and evaluation for non-lambertian and uncalibrated photometric stereo. IEEE Trans Pattern Anal Mach Intell 41(2):271–284
59. Shurcliff W (1962) Polarized light, production and use. Harvard, Cambridge (1962)
60. Smith WA, Ramamoorthi R, Tozza S (2016) Linear depth estimation from an uncalibrated, monocular polarisation image. In: Proceedings of European conference on computer vision, pp 109–125
61. Smith WA, Ramamoorthi R, Tozza S (2019) Height-from-polarisation with unknown lighting or albedo. IEEE Trans Pattern Anal Mach Intell 41(12):2875–2888
62. Stolz C, Ferraton M, Meriaudeau F (2012) Shape from polarization: a method for solving zenithal angle ambiguity. Opt Lett 37(20):4218–4220
63. Taamazyan V, Kadambi A, Raskar R (2016) Shape from mixed polarization. arXiv preprint arXiv:1605.02066
64. Tan Z, Zhao B, Xu X, Fei Z, Luo M (2019) Object segmentation based on refractive index estimated by polarization of specular reflection. Optik, 163918
65. Teo GD, Shi B, Zheng Y, Yeung SK (2018) Self-calibrating polarising radiometric calibration. In: Proceedings of computer vision and pattern recognition
66. Torrance KE, Sparrow EM (1992) Radiometry. Chapter theory for off-specular reflection from roughened surfaces, vol 3. Jones and Bartlett Publishers Inc, USA, pp 32–41
67. Tozza S, Smith WA, Zhu D, Ramamoorthi R, Hancock ER (2017) Linear differential constraints for photo-polarimetric height estimation. In: Proceedings of international conference on computer vision, pp 2279–2287
68. Wieschollek P, Gallo O, Gu J, Kautz J (2018) Separating reflection and transmission images in the wild. In: Proceedings of European conference on computer vision, pp 89–104
69. Wolff LB (1997) Polarization vision: a new sensory approach to image understanding. Image Vis Comput 15(2):81–93
70. Wolff LB, Boult TE (1991) Constraining object features using a polarization reflectance model. IEEE Trans Pattern Anal Mach Intell 13(7):635–657
71. Wolff LB, Nayar SK, Oren M (1998) Improved diffuse reflection models for computer vision. Int J Comput Vis 30(1):55–71
72. Yang L, Tan F, Li A, Cui Z, Furukawa Y, Tan P (2018) Polarimetric dense monocular slam. In: Proceedings of computer vision and pattern recognition, pp 3857–3866
73. Yu Y, Zhu D, Smith W.A (2017) Shape-from-polarisation: A nonlinear least squares approach. In: Proceedings of international conference on computer vision, pp 2969–2976
74. Zheng Q, Jia Y, Shi B, Jiang X, Duan LY, Kot AC (2019) Spline-net: sparse photometric stereo through lighting interpolation and normal estimation networks. In: Proceedings of international conference on computer vision, pp 8549–8558
75. Zhu D, Smith WA (2019) Depth from a polarisation+ RGB stereo pair. In: Proceedings of computer vision and pattern recognition, pp 7586–7595

Chapter 7
Estimating Facial Aging Using Light Scattering Photometry

Hadi A. Dahlan and Edwin R. Hancock

Abstract Facial aging is a complex process, and the changes in the inner layers of the skin will affect how the light scatters from the skin. To observe whether a light scattering model parameter is suitable to be used for age classification/estimation, this study investigated and analyzed the relationship between the parameter of an analytical-based light scattering model and skins of various ages using photometry method. Multiple models are used to investigate and compare the relationship between the model parameters and the subject's age. The results show that all of the models' roughness parameter representation has a significant positive correlation with age ($p < 0.05$), making it a suitable choice to be made as a feature for estimating/classifying age. This study proves that the parameter(s) for an analytical-based light scattering model can be used as an alternative method for estimating/classifying a person's age, provided that we know the light incidence and reflectance angles. In the future, this method can be used to work with other age extractors/estimators/classifiers, for the purpose of designing a more robust age estimation/classification method.

Keywords Age classification · Light scattering model parameter · Photometry · Data analysis · Pattern recognition

7.1 Introduction

Facial aging is an interesting topic that has been studied by various authors from multiple fields. The knowledge gleaned from the studies can be used in numerous real-world settings, such as in plastic surgeries, biometrics, criminology, and even

H. A. Dahlan (✉)
Faculty of Information Science and Technology, The National University
of Malaysia, 43600 UKM Bangi, Selangor, Malaysia
e-mail: had86@ukm.edu.my

E. R. Hancock
Department of Computer Science, University of York, Deramore Lane,
York YO10 5GH, UK
e-mail: edwin.hancock@york.ac.uk

© Springer Nature Switzerland AG 2020
J.-D. Durou et al. (eds.), *Advances in Photometric 3D-Reconstruction*,
Advances in Computer Vision and Pattern Recognition,
https://doi.org/10.1007/978-3-030-51866-0_7

205

animation. Nevertheless, there is still more that we need to learn because facial aging is a very complex process. In most cases, age can be estimated by just observing the changes in face shape and texture [1–3]. However, how does this change affect the skin light scattering? Researchers have discovered that the inner characteristics of the skin, such as the dermis and the epidermis layers, will change with age [4–7], which means that the light scattering on the skin may also change in some way. Can this change be modeled with a light scattering model? And can the model be used as an alternative method to estimate/classify the age?

The aim of this study is to investigate and analyze the light scattering behavior of the facial skin for subjects of different ages, using a light scattering model. This will allow us to explore an alternative method of classifying a person's age using the model parameters and photometry technique. Moreover, to further understand how the aging of the skin affects the light scattering, the model parameters are used as features for age classification. If the study improves further in the future, we possibly can estimate the person's age using the subject's skin reflectance taken from any distance. Moreover, since different ethnicities have a different skin color, we can also further our understanding of the skin aging of different skin colors. The study we are doing here is mainly experimental and also to give some references for the modified light scattering models that we will be using here. Nevertheless, the information obtained from this study regarding model parameters relation with skin aging can be beneficial not only in the computer vision but also in other fields as well, such as pattern recognition, skin optic, or dermatology.

Section 7.2 first explores some facial aging, reflectance and scattering model literature from multiple fields. In Sect. 7.3, the selected models and setup are introduced. Afterward, Sect. 7.4 will be the parameter-age experiments, which are the correlation test between the subject age and the interest parameters, and also the age classification test using the chosen parameters. The Conclusion will be mentioned in the final section.

7.2 Field Survey and Review

7.2.1 Face Aging Research

The innate ability of humans to estimate a person's age based on observations of the facial appearance is impressive. Assigning this ability into a machine capable of learning and applying it to a wide variety of fields, such as in law enforcement, security control, soft-biometric access system, and even in the entertainment industry, has been an interesting and important development. Over the years, multiple facial traits that help to determine a person's age have been identified, including the shape of the face, skin texture, skin features, and skin color contrast [3, 6]. The two predetermined features are: (1) The face shape changing, in particular, the cranium

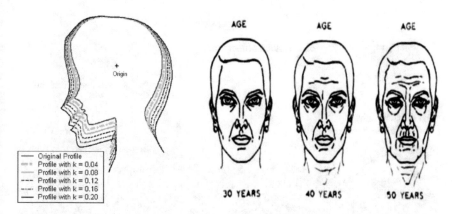

Fig. 7.1 The facial aging in terms of shape or craniofacial growth (left) and texture (right). Images were taken from [1, 6, 8]

bones grow as time passes. This occurs predominately during the transition from childhood to the adult stage, and; (2) The face develops wrinkles (or face texture), as facial muscle wastes due to decreasing elasticity. This occurs during the transition from adulthood to the senior stage [1, 6, 8, 9]. Figure 7.1 shows both of these facial aging features.

As light shines upon the skin, reflectance can be viewed as a result of the combined effect of the optical phenomena induced by the physio-anatomical components of all the skin layers, with each component emitting a specific optical effect [10–12]. As a person grows older, internal and external forces act upon the outer and inner skin causing some level of damage, which later changes the appearance of the skin and its light scattering. This is demonstrated in [4, 5], wherein the young skin was perceived to have a different color contrast and luminosity than the older skin. Healthy, young skin has a smooth, uniformly fine texture that reflects light evenly. It is also plumper and emits radiant color. Meanwhile, aged skin tends to be rough and dry, with more wrinkles, freckles, and age spots, and tends to emit dull color [4–7]. Figure 7.2 for the simple depiction of the young skin layer and the old skin layer.

These discoveries were later applied in the construction of the face aging system in the Computer Vision and Pattern Recognition fields. There are two uses for common facial aging research in the field of computer vision and pattern recognition. They are: (1) face age synthesis [2, 13–17] and (2) face age estimation/classification [1, 8, 18–22]. However, most of these authors did not specifically analyze the light reflectance behavior of facial skin of various ages or thoroughly analyze the effect of aging on the light reflectance/scattering.

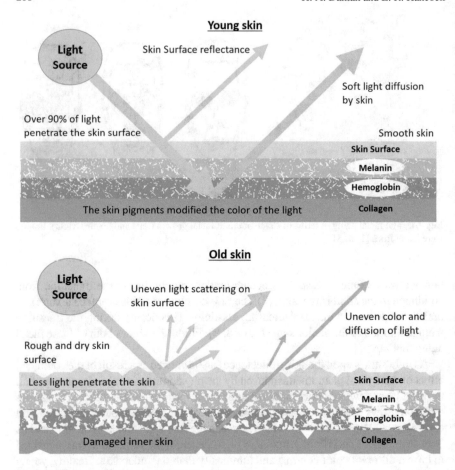

Fig. 7.2 Diagram of light scattering on the skin

7.2.2 Reflectance and Scattering Model

Reflection is a process during which an electromagnetic power flux travels incidentally toward a point of a surface and leaves that point without any change in frequency. The function that is used to model this phenomenon is called the Bidirectional Reflectance Distribution Function (BDRF) [23–25]. The function defines the model that demonstrates how reflected radiance is distributed in terms of the distribution of incident radiance. The light reflection depends on the characteristics of the light, the composition of the material, and its physical traits.

To create the appearance of the skin, the authors introduced the skin light scattering models that account for the layers of the skin [26–30]. Most of these models rely on biophysical parameters from medical and optical tissue literature. Some authors such as Weyrich et al. [31] developed a skin reflectance model wherein the parameters were estimated using multiple measurements that they captured using custom-built devices. While some modeled human skin as a layered component with biophysical-based user parameters and intuitive behavior, such as the melanin and the hemoglobin fraction [32, 33].

Literature in the medical, human perception and skin optical fields have mentioned that the skin characteristics change as the age increases. However, none of the current papers on reflectance/scattering models have thoroughly analyzed how aging affects the reflectance/scattering model predictions or its parameter estimation. We thought it would be interesting if we can study the inner working of the skin in terms of a model-based that can obtain multiple parameter values that have meaning (e.g., the surface roughness and light absorption) and study its relation to the process of aging using both the light scattering model and photometry. So, this paper aims to investigate whether the light scattering model parameters change together with the aging skin, and if so, whether the parameters of a model can be used as a feature for estimating/classifying age.

7.3 Methodology

The main objective of this study is to analyze whether the parameters of statistic-based Bidirectional Reflectance Distribution Function (BRDF) models can be used to distinguish between the skin of differing ages. This is done by first capturing the subject's radiance measurements using a photometry technique. Then, the models are fit to the measurements by varying the model parameters to minimize the root-mean-squared error. Next, correlation is calculated between the parameters and the ages of the subjects to investigate which model parameters are strongly affected by facial aging. Finally, an age classification test is conducted to see how effective the parameters are when used as features for age classification. This section provides details regarding the selected models, how the data was collected and organized, and the methodology of the experiment.

7.3.1 Chosen Models for the Test

When conducting the aging experiment, the face geometry is first estimated using photometric stereo. This is done by acquiring the normal map using the Ma et al. spherical gradient illumination method [34] (more detail on acquisition setup in Sect. 7.3.2). Then a single image is used to estimate the reflectance parameters by fitting a BRDF model to a single input image, knowing the geometry. However, the choice of light scattering model needs to be addressed. The light scattering model parameters must be able to model the characteristics of the skin. These parameters are later estimated using the captured radiance data of the subject's face, provided that we know the four angles, namely, the incident zenith angle θ_i, the scattering zenith angle θ_s, the incident azimuth angle ϕ_i, and the scattering azimuth angle ϕ_s. To estimate the parameter(s), the selected models are fit to the captured radiance data. This was done by varying the model parameters to minimize the root-mean-square error Δ_{RMS}. The RMS fitting error is given by

$$
\Delta_{RMS} = 100 \times \frac{1}{K} \left\{ \sum_{k=1}^{K} \left[L_O^D \left(\theta_i^k, \phi_i^k, \theta_s^k, \phi_s^k \right) \right. \right.
$$
$$
\left. \left. - L_O^P \left(\theta_i^k, \phi_i^k, \theta_s^k, \phi_s^k, param_1, param_2, \ldots, param_c \right) \right]^2 \right\}^{\frac{1}{2}} \quad (7.1)
$$

where L_O^D is the captured radiance data, L_O^P is the radiance from the model prediction, $param_c$ is the model parameter, c is the number of parameters the model has, and k runs over the index number of the BRDF measurements used (K).

For this test, 6 different models were used; two variants of R–H models [35], two variants of A–L models [36], Oren–Nayar model [37], and Jensen model [38]. Each model is having some parameters which can be compared with one another. In this test, we did not include any specular model components.

7.3.1.1 Ragheb–Hancock Model Variants

The first two models are the Ragheb and Hancock (R–H) light scattering model [35]. It is a detailed diffuse light scattering model using the wave scattering theory. The model assumes that the diffuse radiance is scattered from bi-layered rough surfaces, consisting of an opaque subsurface layer below a transparent one. Here, the authors used Vernold and Harvey's version of the Beckmann model for both the surface and subsurface rough scattering effects, while, the Fresnel theory and the Snell's law are used for modeling attenuation factor and the light transmission. Both of the outgoing radiance components (surface and subsurface) are considered identical by the authors. The total outgoing radiance is the linear combination of both components with β as its relative balance control. The R–H model is given as

$$
L_o = \beta L_o^{sb} + (1 - \beta) L_o^{sf} \quad (7.2)
$$

The notations for the R–H model are summarized in Table 7.1.

The R–H model also has two different models L_{o-RH} variants, which are (i) the Gaussian (L_{G-RH}) and (ii) the exponential (L_{E-RH}), in which both refers to the nature of the correlation function for the surface and subsurface roughness. From [35], the scattered surface radiance when the correlation function is Gaussian is given by

Table 7.1 The R–H model formula notation

Notation	Description
L_i	Incident radiance
L_G or L_E	Total scattered radiance (either Gaussian or exponential)
L_G^{sf}	Surface scattered radiance with Gaussian correlation function
L_G^{sb}	Subsurface scattered radiance with Gaussian correlation function
L_E^{sf}	Surface scatter radiance with exponential correlation function
L_E^{sb}	Subsurface scattered radiance with exponential correlation function
θ_i	Surface incident zenith angle
θ_s	Surface scattering zenith angle
θ_i'	Subsurface incident zenith angle
θ_s'	Subsurface scattering zenith angle
ϕ_s	Scattered azimuth angle
σ/T	Surface Root-Mean-Square (RMS) slope
σ'/T'	Subsurface Root-Mean-Square (RMS) slope
K_G or K_E	Coefficients for the surface equations of Gaussian and exponential, respectively
$d\omega'$	Solid angle under mean surface level
n	Standard refractive index
β	Balance parameter

$$L_{G-RH}^{sf}(\theta_i, \theta_s, \phi_s, \sigma/T) =$$

$$K_G\left[\frac{\cos(\theta_i)}{v_z^2(\theta_i, \theta_s)}\right] \times \exp\left[\frac{-T^2 v_{xy}^2(\theta_i, \theta_s, \phi_s)}{4\sigma^2 v_z^2(\theta_i, \theta_s)}\right] \quad (7.3)$$

and when surface correlation function is exponential:

$$L_{E-RH}^{sf}(\theta_i, \theta_s, \phi_s, \sigma/T) =$$

$$K_E\left[\frac{\cos(\theta_i)}{v_z^2(\theta_i, \theta_s)}\right] \times \left(1 + \left[\frac{T^2 v_{xy}^2(\theta_i, \theta_s, \phi_s)}{\sigma^2 v_z^2(\theta_i, \theta_s)}\right]\right)^{-\frac{3}{2}} \quad (7.4)$$

where $v_{xy}^2(\theta_i, \theta_s, \phi_s) = [k(\sin(\theta_i) - \sin(\theta_s)\cos(\phi_s))]^2 + [-k(\sin(\theta_s)\sin(\phi_s))]^2$; $v_z(\theta_i, \theta_s) = -k(\cos(\theta_i) - \cos(\theta_s))$; and $k = 2\pi/\lambda$. The coefficients K_G and K_E are both proportional to $(\sigma/T)^2$ and can be normalized. Meanwhile, the subsurface scattered radiance when the correlation function is Gaussian is given by:

$$L_{G-RH}^{sb}(\theta_i, \theta_s, \phi_s, \sigma'/T', n) =$$

$$L_{G-RH}^{sf}(\theta_i', \theta_s', \phi_s, \sigma'/T') \times [1 - f(\theta_i, n)][1 - f(\theta_s', 1/n)]d\omega' \quad (7.5)$$

and when the subsurface correlation function is exponential:

$$L_{E-RH}^{sb}(\theta_i, \theta_s, \phi_s, \sigma'/T', n) =$$
$$L_{E-RH}^{sf}(\theta_i', \theta_s', \phi_s, \sigma'/T') \times [1 - f(\theta_i, n)][1 - f(\theta_s', 1/n)]d\omega' \quad (7.6)$$

where the subsurface solid angle and the Fresnel coefficient definition can be referred to from [35]. From this model, the parameters which we are interested in the model are the σ/T and β. For comparison with other model surface roughness parameters, the R–H model RMS slope for the surface and the subsurface are made identical $\sigma/T = \sigma'/T'$.

7.3.1.2 Absorption–Light Model Variants

The next two models are the absorption light (A–L) scattering model [36], which is the modified version of the R–H model. The R–H model is detailed but it does not account for light absorption. The A–L model extends the R–H model by adding an absorption term Ab in the subsurface layer; defining it using the conservation of energy for light transmission, reflectance, and absorption. The A–L surface scattering component is similar to those of the R–H surface scattering component for both variants ($L_{G-AL}^{sf} = L_{G-RH}^{sf}$ and $L_{E-AL}^{sf} = L_{E-RH}^{sf}$). However, A–L subsurface scattering component is different, where

$$L_{G-AL}^{sb}(\theta_i, \theta_s, \phi_s, \sigma'/T', n) = L_{G-AL}^{sf}(\theta_i', \theta_s', \phi_s, \sigma'/T')$$
$$\times [1 - f(\theta_i, n) - Ab(a, \theta_i)][1 - f(\theta_s', 1/n) - Ab(a, \theta_{s2})]d\omega' \quad (7.7)$$

$$L_{E-AL}^{sb}(\theta_i, \theta_s, \phi_s, \sigma'/T', n) = L_{E-AL}^{sf}(\theta_i', \theta_s', \phi_s, \sigma'/T')$$
$$\times [1 - f(\theta_i, n) - Ab(a, \theta_i)][1 - f(\theta_s', 1/n) - Ab(a, \theta_{s2})]d\omega' \quad (7.8)$$

where

$$\theta_{s2} = \sin^{-1}\left[\frac{\sin(\theta_s'))}{1/n}\right] \quad (7.9)$$

where a is the fractional absorption parameter, used to control how strongly light is absorbed. A more detailed description of the model can be referred to [36]. The parameters which we are interested in for these models are the σ/T, a, and β. Similarly, this model RMS slope for the surface and the subsurface are also made identical $\sigma/T = \sigma'/T'$. Meanwhile, n is the refractive index for the models. The refractive index n was not included as the chosen parameter because when initially tested on the model parameter estimation range (more in Sect. 7.4), the n estimated value of the samples tend to highly deviate from the reference values of skin ($n = 1.37$ [35]). Hence, for this experiment, the n for the models were fixed to 1.37.

7.3.1.3 Oren–Nayar Model

The fifth model is the Oren–Nayar model, which is a comprehensive model for surfaces with macroscopic roughness. It was initially derived for anisotropic surfaces that have one slope facets and later used for developing a model for isotropic surfaces with Gaussian slope-area distribution. The equation for this model was based on [37] equation derivation, which includes both the direct illumination component and the inter-reflections component. The model radiance equation is

$$
L_{ON_1}(\theta_i, \theta_s, \phi_s - \phi_i, m) = \rho_{ON}/\pi E_0 \cos\theta_i [C_1(m) \\
+ \cos(\phi_s - \phi_i)C_2(\nu_{max}; \nu_{min}; \phi_s - \phi_i; m)\tan(\nu_{min}) \\
+ (1 - |\cos(\phi_s - \phi_i)|)C_3(\nu_{max}; \nu_{min}; m)\tan\left(\frac{\nu_{max} + \nu_{min}}{2}\right)] \quad (7.10)
$$

$$
L_{ON_2} = \frac{0.17\rho_{ON}^2 m^2}{\pi(m^2 + 0.13)} E_0 \cos\theta_i \left[1 - |\cos(\phi_s - \phi_i)|\left(\frac{2\nu_{min}}{\pi}\right)^2\right] \quad (7.11)
$$

where ρ_{ON} is the model albedo, $E_0 \cos\theta_i$ is the irradiance where it is assumed here $E_0 = 1$, m is the model surface roughness measured in degrees, $\nu_{max} = max[\theta_s, \theta_i]$, and $\nu_{min} = min[\theta_s, \theta_i]$. The coefficients of the model are

$$
C_1 = 1 - 0.5\frac{m^2}{m^2 + 0.33} \quad (7.12)
$$

$$
C_2 = \begin{cases} 0.45\frac{m^2}{m^2+0.09}\sin\nu_{max} & if \cos(\phi_s - \phi_i) \geq 0, \\ 0.45\frac{m^2}{m^2+0.09}(\sin\nu_{max} - (\frac{2\nu_{min}}{\pi})^3) & otherwise \end{cases} \quad (7.13)
$$

$$
C_3 = 0.125\frac{m^2}{m^2 + 0.09}\left(\frac{4\nu_{max}\nu_{min}}{\pi^2}\right)^2 \quad (7.14)
$$

The total radiance for the Oren–Nayar is

$$
L_{ON} = (L_{ON_1} + L_{ON_2})k_d \quad (7.15)
$$

where k_d is the diffuse component. The Oren–Nayar model has only one parameter that we interest in, which is the roughness parameter m.

7.3.1.4 Jensen Model

The final model selection is the Jensen model [38]. It is a model that combines a dipole diffusion approximation with single scattering computation. The intention of the model creation was to create a BRDF model that considers subsurface scattering in translucent materials. In this experiment, we used Jensen's BSSRDF approximation total diffuse reflectance equation, which is given by

$$L_J = \frac{\alpha'}{2\pi}\left(1 + \exp^{-\frac{4}{3}\frac{1+Fr}{1-Fr}\sqrt{3(1-\alpha')}}\right)\exp^{-\sqrt{3(1-\alpha')}} \tag{7.16}$$

where Fr is the Fresnel formula and α' is the apparent albedo, given by $\alpha' = \sigma_s/\sigma_s + \sigma_a$. Here, σ_s and σ_a are both scattering and absorption coefficient, which are both the parameters we are interested in. In the parameter-age test, the parameters that we are interested in will be estimated in a specific range. More explanation on this will be mentioned in Sect. 7.4.

7.3.2 The Data Acquisition and Organization

In this experiment, two measurements were taken from the subject; the normal map and the subject's radiance, both captured in a frontal position. The normals are used for deducing the zenithal and azimuthal angles, while the radiance image is used as a reference image for the model fitting. The normal map was acquired using the Ma et al. spherical gradient illumination method [34] using a geodesic dome light stage with a diameter of 1.58 m, with 40 filtered LEDs located at each geodesic vertex [39]. Linear polarization filters were positioned in front of each of the LEDs and also in front of the lens of the Nikon D200 camera. The light stage is controlled using an mbed NXP LPC1768 microcontroller development board. This controls the brightness of the LEDs and the camera shutter. The camera was placed at the periphery of the light stage and was manually focused. A single-camera flash lamp (Nikon Speed-light SB-600) was positioned slightly above the camera, having a flash output level of 1/64 and a zoom head position value configured to 24 mm. The subjects in this experiment were volunteers, all of whom had freely provided their consent. The candidates were categorized based on their skin types using the Fitzpatrick labeling method, which is a skin classification based on the amount of melanin pigment in the skin [40]. Some volunteers were wearing makeup (or facial cream) when the image of them was captured. It should be noted that cosmetics do likely affect the light scatter on the skin. More than half of the women in the dataset were wearing cosmetics when the image of them was captured. In the experiment, these volunteers are included in the test to observe how would the cosmetics affect the estimation value of the model parameter(s). Table 7.2 shows the candidate pool grouped based on their gender and skin type.

Figure 7.3 shows the geodesic light stage used for this experiment and its diagram. A single diffuse reflectance image of the face was captured in a frontal position using a single-flash camera fitted with a cross-polarizer filter. The images were captured in the .tiff format with the size being 3900×2616, however, for processing, the images were resized to $1/4$ (or 975×654) of its original size using the bicubic interpolation method. The scaled-down image reduces the time required for the image alignment

Table 7.2 The candidates used for the parameter-age test

Group by gender

Age group	19–29	30–39	40–49	50>	Total
All male	9	7	1	6	23
All female	14	6	4	1	25
Overall total					48

Group by skin type (using the Fitzpatrick labeling method [40])

Skin type	Male	Male with cream	Female	Female with makeup/cream	Total
2	4	0	2	3	9
3	9	1	3	5	18
4	6	0	6	6	18
5	2	1	0	0	3
Overall total					48

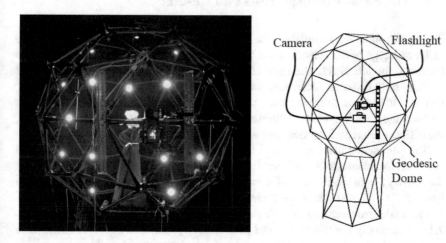

Fig. 7.3 The geodesic dome light stage used in the experiment

process. The authors have noted that reducing the image size will degrade the image quality to some extent due to the down-sampling process, which will affect the appearance quality of the aging features that are on the facial image. Figure 7.4 shows some of the volunteers captured using the camera flash fitted with a linear polarization filter. The details for the setup image processing can be referred to in [41].

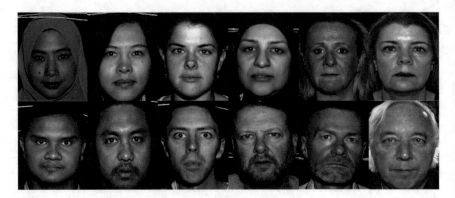

Fig. 7.4 Several images of volunteers used in this experiment

7.4   The Parameter-Age Tests and Results

Due to the small data available, the images that had been captured, which is the radiance image of the subject, are cropped into four different sections, namely, the forehead, the left cheek, the right cheek, and the nose. Since the different facial sections have different facial structures and skin properties that produce differing radiance intensities, cropping each subject images into 4 different sections helps increase the overall data count (meaning four times more data). In this experiment, the radiance from each section area was averaged across the angles.

The parameters that are selected for this test are all the parameters that should appear as parameters to the radiance functions from their respective models. Meaning that the σ/T and β are both referring to the R–H and A–L models parameters in Eqs. 7.5, 7.6, 7.7, and 7.8; the A–L model parameter a in Eqs. 7.7, and 7.8; the Oren–Nayar model parameter m in their radiance function equation (referred Eqs. 7.10–7.15); and finally the Jensen's model parameters σ_s and σ_a in Eq. 7.16. To estimate the parameter(s), the selected models are fit to the captured radiance data. This was done by varying the model parameters to minimize the root-mean-square error, Δ_{RMS} (see Eq. 7.1). For the A–L model variants (exponential and Gaussian), the initial estimation range for its RMS slope σ/T is between (0.01 and 2.00), while the absorption a—range between (0 and 1). Meanwhile, for the Ragheb–Hancock (R–H) models, σ/T is between (0.01 and 4.00). The balance parameter β for both the A–L and R–H models is made within the range of (0.01 and 1.00). It should be noted that in this test, for both of these models, the RMS slope for the surface and the subsurface are made identical $\sigma/T = \sigma'/T'$. As for the Oren–Nayar model, the roughness parameter m is measured in degrees, so the range of estimation is between (0° and 180°); while the albedo ρ_{ON} is within (0–1). The BSSRDF model scattering σ_s and absorption σ_a coefficients were also estimated to be between (0 and 1). Each model has a coefficient k_d diffuse component, in which the range for them being between (0 and 1).

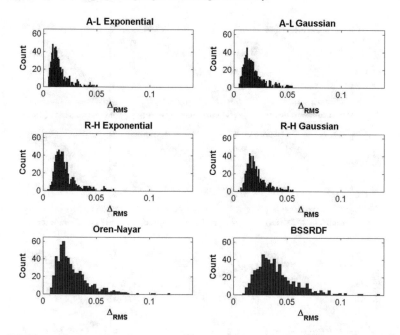

Fig. 7.5 The root-mean-square error RMS_Δ histogram for the models using all samples (including all facial region)

The estimations are completed using 1000 selected sample points, using MAT-LAB's least-squared function. After the initial fitting process, the samples are fitted to the model again, but with smaller and more specific estimation ranges for the parameters. This is done to decrease large variations within the initial estimation. The process is completed several times until it reaches the least total Δ_{RMS} and the least possible variation. To demonstrate the accuracy of the models, Fig. 7.5 shows the Δ_{RMS} histogram of all the data, including all of the color channels and all of the face sections for all six different models. The subjects used in this experiment possessed a wide variety of facial features as a result of varying gender, skin type, cosmetic use, and facial structure.

To initially determine which of the models investigated perform the best fitting with the data measurements, we use scatter plots to observe how closely the data prediction of the model to that of the data measurement. Figure 7.6 illustrates the scatter plot for normalized radiance measurements against the data prediction from six different models done on one subject and for each RGB channel. From the plot, the better the data is clustered around the diagonal straight line, the better the agreement between experiment and theory. It was observed that the A–L model for the exponential variant gives the radiance prediction closest to that of the data measurement for all color channels. Meanwhile, Figs. 7.7, 7.8, 7.9, 7.10, 7.11, and 7.12 show the plots for the subject age against the estimated parameters for each of the six models, on each RGB channel. The experiment used all the subjects; including all

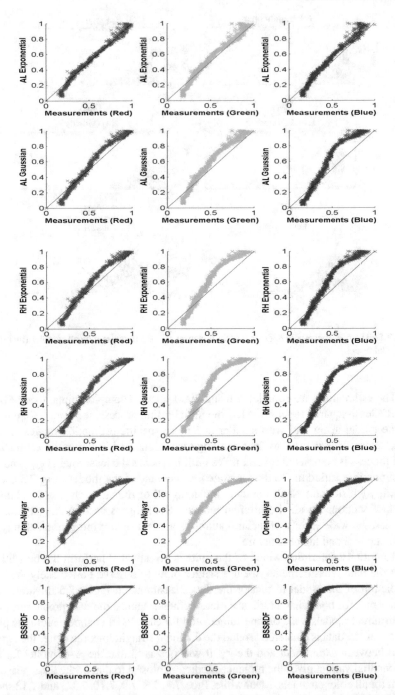

Fig. 7.6 Normalized data measurements against the normalized radiance prediction (Subject: Male, Age: 35, Skin Type: 2, Face Section: Forehead, Cosmetic: None)

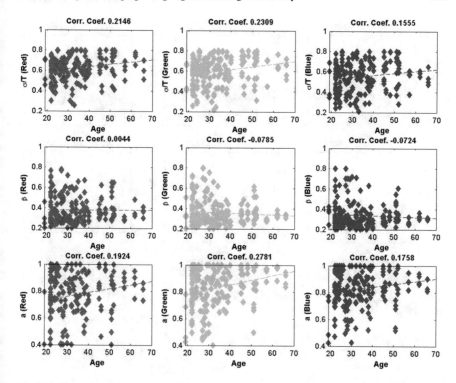

Fig. 7.7 The estimated parameter plots for the A–L exponential model done using all subjects; including all facial sections. The top of each plot shows its correlation coefficient

facial sections, giving a total of 192 samples. The points in the plots are the estimated model parameters. The dotted line is the regression line. Next, Tables 7.3, 7.4, and 7.5 show the two-tailed correlation coefficient results between each model parameters and the subject age.

7.4.1   The Parameter-Age Correlation Test Results, Analysis, and Discussion

In, the search for the best-fit parameters, the A–L model exponential has the lowest Δ_{RMS}, followed by the A–L model Gaussian. The BSSRDF exhibited the overall highest error, followed by the Oren–Nayar model (see Fig. 7.5). The BSSRDF and the Oren–Nayar models were both designed to model large diffuse radiance, so, in this test, they failed to properly fit to sample radiances that are less diffuse. Meanwhile, the σ/T and β estimated values for the A–L models are lower than the R–H model counterpart (Fig. 7.7, 7.8, 7.9 and 7.10). This is due to the parameter a increasing the flexibility of the A–L model. Moreover from the graphs Figs. 7.7, 7.8, 7.9, and

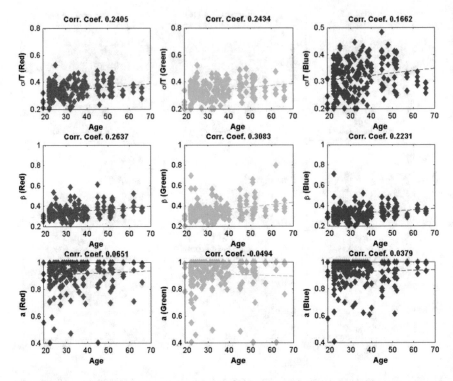

Fig. 7.8 The estimated parameter plots for the A–L Gaussian model done using all subjects; including all facial sections. The top of each plot shows its correlation coefficient

7.10, saturation to 0 or 1 can be seen. Due to the small data, it is difficult to consider them as outliers. These data are still included in the graph because they represent the undersampled part of the data (subjects $\geq$ 40 old) and removing them will cause the model only represents part of the real data and cannot predict a significant portion of reality.

7.4.1.1   Results Analysis and Discussion

To recall the definition of the p-value in statistics; the p-value is the evidence against a null hypothesis (randomness). In other words, the smaller the p-value (e.g., $p < 0.1$), the smaller chance your results could be random or happened by chance. Looking at the parameter-age test results. An interesting pattern emerged from the correlation results when all samples were considered (Tables 7.3, 7.4, and 7.5). The roughness parameters for the A–L, R–H, Oren-Nayar, and BSSRDF models (e.g., σ/T, m, and σ_s) suggest that a significant positive correlation with age exists ($p < 0.05$). This is true for both male and female subjects, including the subjects that wear makeup or facial cream. These results support the literature regarding skin roughness. However,

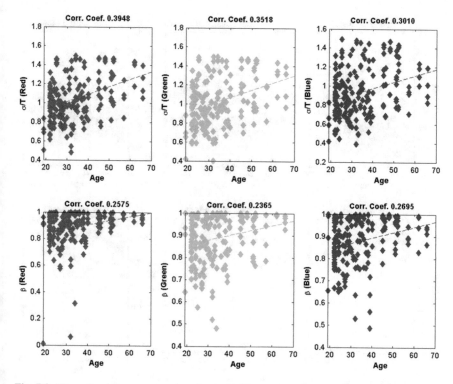

Fig. 7.9 The estimated parameter plots for the R–H exponential model done using all subjects; including all facial sections. The top of each plot shows its correlation coefficient

the β, a, and σ_a parameters show no strong pattern or relation with age. This is because these parameters are more affected by skin type and condition. The A–L model exponential parameters have more variation for the a parameter than the Gaussian version. However, the Gaussian version has a higher error count than the exponential one. The regression line is fitted on the estimated parameter to observe the relation between the age and the parameters. The chi-square for fit calculated was ≈ 1.008 for all the plots (see Figs. 7.7, 7.8, 7.9, 7.10, 7.11, and 7.12). Meanwhile, the Oren–Nayar and BSSRDF model parameters have outliers that affect the regression line placement (see Figs. 7.11 and 7.12).

7.4.2 The Roughness Parameter-Age Correlation Test for a Specific Category

Multiple studies have suggested that factors like gender, ethnicity, and cosmetic use do result in differences in appearances as people age chronologically [4, 5, 42–48]. As a result of this, we have undertaken some additional parameter-age tests based

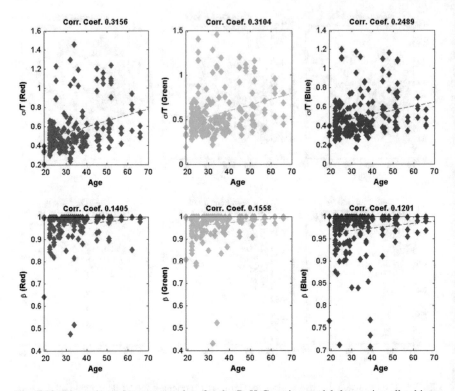

Fig. 7.10 The estimated parameter plots for the R–H Gaussian model done using all subjects; including all facial sections. The top of each plot shows its correlation coefficient

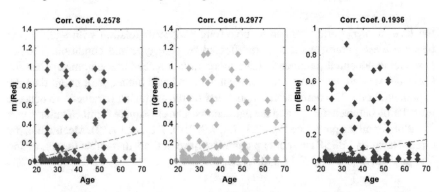

Fig. 7.11 The estimated parameter plot for the Oren–Nayar model done using all subjects; including all facial sections. The top of each plot shows its correlation coefficient

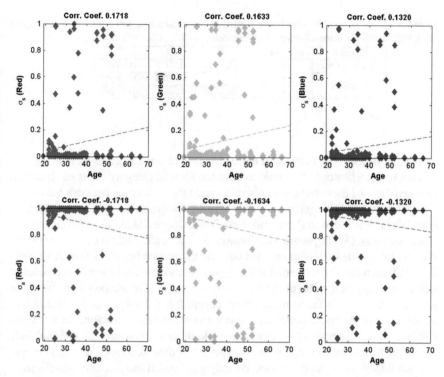

Fig. 7.12 The estimated parameter plots for the BSSRDF done using all subjects; including all facial sections. The top of each plot shows its correlation coefficient

Table 7.3 Correlation coefficients between the parameter and the subject age for the A–L model using all samples. (White cell $p < 0.05$; Gray cell: $p > 0.2$)

Model	Parameter	R	G	B
	σ/T	0.2146	0.2309	0.1555
A–L Exponential	β	0.0044	-0.0785	-0.0724
	a	0.1924	0.2781	0.1758
	σ/T	0.2405	0.2434	0.1662
A–L Gaussian	β	0.2637	0.3083	0.2231
	a	0.0651	-0.0494	0.0379

Table 7.4 Correlation coefficients between the parameter and the subject age for the R–H model using all samples. (White cell $p < 0.05$; Gray cell: $0.1 < p < 0.2$)

Model	Parameter	R	G	B
R–H Exponential	σ/T	0.3948	0.3518	0.3010
	β	0.2575	0.2365	0.2695
R–H Gaussian	σ/T	0.3156	0.3104	0.2489
	β	0.1405	0.1558	0.1201

Table 7.5 Correlation coefficients between the parameter and the subject age for the Oren–Nayar and BSSRDF models using all samples. (White cell $p < 0.05$; Gray cell: $0.1 < p < 0.2$)

Model	Parameter	R	G	B
Oren–Nayar	m	0.2578	0.2977	0.1936
BSSRDF	σ_s	0.1718	0.1633	0.1320
	σ_a	-0.1718	-0.1634	-0.1320

on specific factors. The purpose of these tests is to observe whether there is any significant result when only the subjects with the specific category are used. It should be noted that, for these tests, we only considered the correlation between the roughness parameter and the age of the subject. These categories are (1) Gender (Male or Female); (2) Face section (Forehead, Left Cheek, Right Cheek, Both Cheeks, Nose, or All sections); (3) Fitzpatrick Skin Type (2, 3, 4, 5, or all); and (4) Cosmetics (With, Without or include both with and without). As mentioned before in Sect. 7.3.2, subjects with cosmetics are included in the experiment to observe how would cosmetics affect the estimation value of the model parameter(s). The methodology for these parameter-age tests is the same as above, except that the subject in the dataset that fits with one of the chosen categories was selected for the experiment. For example, if the chosen category was to use only the male subjects, then only the male subjects were used for the parameter-age test, regardless of their skin type, face section, and whether they wore cosmetics or not. We only highlighted and discussed results from the categories which have the most significant correlation that appears on most of the models. These highlighted categories are the gender (male and female) and face sections (forehead and both left and right cheek). Other results may be mentioned in the circumstance where they warrant attention.

7.4.2.1 The Roughness Parameter-Age Results, Analysis, and Discussion

Tables 7.6 and 7.7 show the roughness parameters and age correlation results within the model for males and females in each gender category. Comparing the two tables (all male versus all female), the results show that most of the light scattering models obtained a positive significant correlation only when the female subjects were used. This result seems to imply that the age of the female candidates is able to be distinguished more easily than that of the males and that the results from the female candidates can be more easily compared when the model roughness parameter is used for the age estimation. However, half of the female candidates wore cosmetics, and, as outline above, cosmetics tend to affect the light scattering, which will likely affect their overall skin light scattering. The purpose of makeup or facial cream is to make skin scatter light evenly [4–7]. However, interestingly enough, since most females in the samples across the age groups wore makeup, the results still show that the older female has rougher skin than the younger ones. This indicates that the

Table 7.6 Correlation coefficients between the roughness parameter and the subject age for all models using just the male candidates. (Gray cell $p < 0.05$; White cell: $0.1 < p < 0.2$)

Model	Roughness Parameter	R	G	B
A-L Exponential	σ/T	0.2204	0.2345	0.1756
A-L Gaussian	σ/T	-0.0096	0.05847	-0.0585
R-H Exponential	σ/T	0.3516	0.2852	0.2374
R-H Gaussian	σ/T	0.1944	0.2016	0.1476
Oren-Nayar	m	0.1730	0.2426	0.1107
BSSRDF	σ_s	0.0217	0.0155	0.0023

Table 7.7 Correlation coefficients between the roughness parameter and the subject age for all models using just the female candidates. (Grey cell $p < 0.05$; White cell: $0.1 < p < 0.2$)

Model	Roughness Parameter	R	G	B
A-L Exponential	σ/T	0.1087	0.1492	0.0730
A-L Gaussian	σ/T	0.3767	0.32354	0.2743
R-H Exponential	σ/T	0.3347	0.3188	0.2731
R-H Gaussian	σ/T	0.3682	0.3481	0.2832
Oren-Nayar	m	0.3124	0.3013	0.2702
BSSRDF	σ_s	0.3779	0.3514	0.3167

older they are, the more makeup (or cream) they apply on their face, hence, the more light scatters due to the multiple layers of makeup they have.

Meanwhile, Tables 7.8 and 7.9 show the correlation results between the model roughness parameters and age for the Forehead section and Both Cheek sections in the face section category. Given that most of the models with strong positive correlations occur within these sections, it may be implied that these sections are the best sections for estimating age using the roughness parameter. Meanwhile, the correlation between the skin type and age, and the correlation between the cosmetics categories (With, Without, or include both with and without) and age was observed to have mostly weak correlation results; very few models are shown to have significant correlation results. This is because when the dataset samples are divided based on these categories, the distribution of the subjects across the various ages becomes uneven, which may, in turn, affect the correlation process. Lastly, even though there are some correlation results in Tables 7.6, 7.7, 7.8, and 7.9 having $(0.1 < p < 0.2)$, the conclusion on the roughness parameter strong relation with skin aging was based on the majority of the results having significant correlation results ($p < 0.05$).

Table 7.8 Correlation coefficients between the roughness parameter and the subject age for all models using just the forehead section of all candidates. (Gray cell $p < 0.05$; White cell: $0.1 < p < 0.2$)

Model	Roughness Parameter	R	G	B
A-L Exponential	σ/T	0.3178	0.4057	0.3649
A-L Gaussian	σ/T	0.3967	0.5033	0.4154
R-H Exponential	σ/T	0.5502	0.5651	0.5658
R-H Gaussian	σ/T	0.4698	0.5430	0.5379
Oren-Nayar	m	0.3688	0.4847	0.3702
BSSRDF	σ_s	0.1700	0.2276	0.1431

Table 7.9 Correlation coefficients between the roughness parameter and the subject age for all models using both left and right cheek sections of all candidates. (Gray cell $p < 0.05$; White cell: $0.1 < p < 0.2$)

Model	Roughness Parameter	R	G	B
A-L Exponential	σ/T	0.2554	0.2291	0.0631
A-L Gaussian	σ/T	0.2627	0.2239	0.1195
R-H Exponential	σ/T	0.4313	0.3650	0.2887
R-H Gaussian	σ/T	0.3510	0.3304	0.2335
Oren-Nayar	m	0.3256	0.3528	0.2287
BSSRDF	σ_s	0.2385	0.2016	0.1732

7.4.3 Age Classification Using the Model Parameters

To test how the parameter fare when using as an aging feature, an age classification experiment was conducted using all the subjects, which include all gender, skin type, and face sections, giving a total of 192 samples overall. The method for classification was done using the support vector machine (SVM) [49], which is a supervised machine learning algorithm that can analyze data to be used for classification.

The experiment is done by first selecting a parameter as the aging feature (i.e., roughness), then hold-out 10% of the samples to be used as testing samples, while the rest as training samples. The tests are done for each model parameter. In this experiment, the samples are group into their specific age groups: (1) Subjects below age 30 as age group 20, (2) Subjects age between 30 and 39 as age group 30, (3) Subjects age between 40 and 49 as age group 40, and (4) Subjects above age 49 as age group 50. The experiment is done on different color channels to observe if there any difference in accuracy between channels. Here, the age classification is done by classifying the test subjects between two different age groups; for example, classify subjects between age group 20 versus 30; 20 versus 40; 20 versus 50; 30 versus 40; 30 versus 50; and 40 versus 50. The average accuracy is calculated based on the correct classification of the subject with its true age.

Table 7.10 The age estimation accuracy for all samples (gender, skin types, and face sections) using the model roughness parameters as the aging feature. This table for the A–L models (both exponential and Gaussian variants)

Parameter (Rough-ness)	A–L exponential ($\frac{\sigma}{T}$)			A–L Gaussian ($\frac{\sigma}{T}$)		
	R	G	B	R	G	B
Accuracy (%)	70.13	69.24	69.24	69.67	69.67	68.48
Total average accuracy (%)	69.54			69.28		

Table 7.11 The age estimation accuracy for all samples (gender, skin types, and face sections) using the model roughness parameters as the aging feature. This table for the R–H model (both exponential and Gaussian variants)

Parameter (roughness)	R–H exponential ($\frac{\sigma}{T}$)			R–H Gaussian ($\frac{\sigma}{T}$)		
	R	G	B	R	G	B
Accuracy (%)	66.9	67.24	69.08	67.59	68.63	68.48
Total average accuracy (%)	67.74			68.24		

Tables 7.10, 7.11, 7.12, 7.13, 7.14, 7.15, and 7.16 show the age classification accuracy results for each of the different model parameters used as the aging feature. From the results, it can be seen that all of the model parameters give a classification accuracy of $\approx$70%, but among all of the models, the parameters estimated using the A–L exponential model are the parameters that can give the best age classification when used as the aging feature. However, when classifying age, the classification must be based on the confirmation that the chosen parameter has a strong relation with skin age progression. Among the selection of parameters available (e.g., σ/T, β, a, m, $sigma_a$, and $sigma_s$), the roughness parameter was previously shown to have the strongest relation with skin aging (the previous correlation tests on parameters). Hence, if biology skin aging progression is considered for future age classification, then it is best to use the model's roughness parameter.

Regarding the test on color channels. Literature in skin optic and computer graphic [30, 50] mentioned that the blue light (wavelength $\approx$450–485 nm) and the green light (wavelength $\approx$500–565 nm) are mostly absorbed by the inner skin while the red light (wavelength $\approx$625–740 nm) is mostly reflected. When the subject becomes older, the skin tends to become less effective in absorbing all three of the color lights (e.g.,

Table 7.12 The age estimation accuracy for all samples (gender, skin types, and face sections) using the model roughness parameters as the aging feature. This table is for the Oren–Nayar model and the BSSRDF model

Parameter (roughness)	Oren–Nayar (m)			BSSRDF (σ_s)		
	R	G	B	R	G	B
Accuracy (%)	68.04	68.63	65.81	67.89	67.29	66.25
Total average accuracy (%)	67.5			67.14		

Table 7.13 The age estimation accuracy for all samples (gender, skin types, and face sections) using the model balance parameters as the aging feature. This table for the A–L models (both exponential and Gaussian variants)

Parameter (balance parameter)	A–L exponential (β)			A–L Gaussian (β)		
	R	G	B	R	G	B
Accuracy (%)	69.74	71.33	69.24	67.20	67.94	67.16
Total average accuracy (%)	69.94			67.43		

Table 7.14 The age estimation accuracy for all samples (gender, skin types, and face sections) using the model balance parameters as the aging feature. This table for the R–H model (both exponential and Gaussian variants)

Parameter (balance parameter)	R–H exponential (β)			R–H Gaussian (β)		
	R	G	B	R	G	B
Accuracy (%)	67.16	67.16	69.84	67.16	69.24	68.65
Total average accuracy (%)	68.05			68.35		

Table 7.15 The age estimation accuracy for all samples (gender, skin types, and face sections) using the model absorption parameters as the aging feature. This table for the A–L models (both exponential and Gaussian variants)

Parameter (absorption)	A–L exponential (a)			A–L Gaussian (a)		
	R	G	B	R	G	B
Accuracy (%)	69.24	69.24	69.24	69.24	67.71	67.16
Total average accuracy (%)	69.24			68.03		

Table 7.16 The age estimation accuracy for all samples (gender, skin types, and face sections) using the model absorption parameters as the aging feature. This table is for the BSSRDF model

Parameter (absorption)	BSSRDF (σ_a)		
	R	G	B
Accuracy (%)	67.89	67.29	66.25
Total average accuracy (%)	67.14		

older skin reflected more red, green, and blue light than the younger skin). In the test results between the three color channels (RGB), the total average accuracy for each channel are $Red = 68.30\%$, $Green = 68.51\%$, and $Red = 68.07\%$. These results seem to imply that the age classification is best done on the green channel. However, from these results, it is still difficult to confirm which color light (RGB) is the most affected by the skin condition. Meanwhile, in this age classification experiment, there is a problem classifying the subjects that are above age group 20 into their corrected age group (not shown here). This may be due to the uneven distribution of subjects between age groups, where, subjects in the age group 20 having more than the other age groups. To solve this problem in the future, adding more subjects with age above 30 years old may improve the SVM training process of this experiment. Another alternative is by solving the uneven data distribution problem using weighted examples.

7.4.4 Limitations and Future Studies

The idea of using the light scattering of the skin to estimate age is a novel idea. However, when conducting the experiments, there were some limitations encountered. The first main limitation is the limited subjects (or volunteers) collected for the main parameter-age test. If there were more subjects, especially for subjects aging above 30 years old, the test results will be more concrete. Older candidates exhibited difficulty entering the light stage since the light stage has no proper door through which they could enter. The physical condition and comfort of the volunteers should be considered more carefully in the future. The authors will also take note of the importance of consent, especially female subjects when requiring them to remove their makeup in the future. The second main limitation is that there are no real measurements for the estimated parameters to compare with. The estimated parameters were assumed to be true based on the model fitting to the subject's radiance image. More specialized tools are required for this kind of problem; collaboration with the researcher in the field of skin optic may also be needed. Finally, this study focused on estimating age using the radiance from the facial skin. This study may also work if another skin area is used instead, such as the radiance from the subject's hand. However, the facial skin area was chosen because it is the skin area that has the most consistent exposure to the sun (photoaging) for most people compared to other skin areas. Nevertheless, the idea of doing this experiment using other skin areas beside the face, such as the hand, is untried yet and conducting it in the future study may give us a new discovery.

7.5   Conclusion

We have provided details regarding an experiment designed to determine whether the light scattering models estimated parameters correlate with the age of various subjects. The behavior of the roughness parameters (σ/T, m, and σ_s) and the parameters of all models are in line with the skin optic and computer vision literature; that is to say that the roughness parameter (or skin surface/subsurface roughness) increases with age. The results in this chapter suggest that the selected light scattering models, especially the A–L models (exponential and Gaussian variants), can be used to approximately determine the age of a subject. This is proven in the age classification experiment (in Sect. 7.4.3) where the model parameters are used as the aging feature for training and testing; while the support vector machine is used as the classifier. Meanwhile, light scatter differs across age are more apparent on the forehead section and both the left and right cheeks (when included together). This is understandable since those sections are where aging features tend to appear. However, it should be noted that the confirmation of the results of the experiment regarding the relationship between the model parameter and the subject age would be much stronger if a dataset that featured similar subjects captured as they age progressively was considered, instead of using different subjects with different ages. Nevertheless, this study suggests that while using photometry, the parameters of an analytical-based light scattering model can be used to estimate a person's age. Moreover, it is possible to use this alternative method in conjunction with another age extractor/estimator method to increase the accuracy of the estimate.

References

1. Fu Y, Guo G, Huang TS (2010) Age synthesis and estimation via faces: a survey. IEEE Trans Pattern Anal Mach Intell 32(11):1955–1976
2. Ramanathan N, Chellappa R (2006) Modeling age progression in young faces. In: 2006 IEEE computer society conference on computer vision and pattern recognition, vol 1. IEEE
3. Rhodes MG (2009) Age estimation of faces: a review. Appl Cognit Psychol 23(1):1–12
4. Fink B et al (2011) Differences in visual perception of age and attractiveness of female facial and body skin. Inter J Cosmetic Sci 33(2): 26–131
5. Fink B, Grammer K, Matts PJ (2006) Visible skin color distribution plays a role in the perception of age, attractiveness, and health in female faces. Evol Human Behav 27(6):433–442
6. Zimbler MS, Kokoska MS, Regan Thomas J (2001) Anatomy and pathophysiology of facial aging. Facial Plastic Surg Clin North Am 9(2):179–187
7. Igarashi T, Nishino K, Nayar SK (2007) The appearance of human skin: a survey. Found Trends Comput Graph Vis 3(1):1–95
8. Ramanathan N, Chellappa R, Biswas S (2009) Age progression in human faces: a survey. J Vis Lang Comput 15:3349–3361
9. Ramanathan N, Chellappa R, Biswas S (2009) Computational methods for modeling facial aging: a survey. J Vis Lang Comput 20(3):131–144
10. Anderson RR, Parrish JA (1981) The optics of human skin. J Invest Dermatol 77(1):13–19
11. Anderson RR, Parrish JA (1982) Optical properties of human skin. In: The science of photomedicine. Springer, Boston, pp 147–194
12. Neerken S, et al (2004) Characterization of age-related effects in human skin: a comparative study that applies confocal laser scanning microscopy and optical coherence tomography. J Biomed Optics 9(2):274–282
13. O'Toole AJ, et al (1999) 3D shape and 2D surface textures of human faces: the role of "averages" in attractiveness and age. Image Visi Comput 18(1):9–19
14. Liu Z, Zhang Z, Shan Y (2004) Image-based surface detail transfer. IEEE Comput Graph Appl 24(3):30–35
15. Vetter T (1998) Synthesis of novel views from a single face image. Int J Comput Vis 28(2):103–116
16. Golovinskiy A, et al (2006) A statistical model for synthesis of detailed facial geometry. ACM Trans Graph (TOG) 25(3):1025–1034
17. Patterson E, et al (2007) Aspects of age variation in facial morphology affecting biometrics. In: BTAS 2007. First IEEE international conference on biometrics: theory, applications, and systems, 2007. IEEE
18. Kwon YH, da Vitoria Lobo N (1999) Age classification from facial images. Comput Vis Image Underst 74(1):1–21
19. Geng X, Zhou Z-H, Smith-Miles K (2007) Automatic age estimation based on facial aging patterns. IEEE Trans Pattern Anal Mach Intell 29(12):2234–2240
20. Fu Y, Xu Y, Huang TS (2007) Estimating human age by manifold analysis of face pictures and regression on aging features. In: 2007 IEEE international conference on multimedia and expo. IEEE
21. Choi SE, et al (2011) Age estimation using a hierarchical classifier based on global and local facial features. Pattern Recognit 44(6):1262–1281
22. Han H, et al (2015) Demographic estimation from face images: human vs. machine performance. IEEE Trans Pattern Anal Mach Intell 37(6):1148–1161
23. Montes Soldado R, Ureña Almagro C (2012) An overview of BRDF models
24. Forsyth DA, Ponce J (2012) Computer vision: a modern approach, 2nd edn. Pearson, London
25. Wynn C (2000) An introduction to BRDF-based lighting. Nvidia Corporation
26. Hanrahan P, Krueger W (1993) Reflection from layered surfaces due to subsurface scattering. In: Proceedings of the 20th annual conference on computer graphics and interactive techniques. ACM

27. Koenderink J, Pont S (2003) The secret of velvety skin. Mach Vis Appl 14(4):260–268
28. Tsumura N, et al (2003) Image-based skin color and texture analysis/synthesis by extracting hemoglobin and melanin information in the skin. ACM Trans Graph (TOG) 22(3):770–779
29. Krishnaswamy A, Baranoski GVG (2004) A biophysically-based spectral model of light interaction with human skin. In: Computer graphics forum, vol 23. No. 3. Blackwell Publishing, Inc, Malden
30. Iglesias–Guitian JA, et al (2015) A biophysically-based model of the optical properties of skin aging. Comput Graph Forum 34(2)
31. Weyrich T, et al (2006) Analysis of human faces using a measurement-based skin reflectance model. ACM Trans Graph (TOG). vol 25. No. 3. ACM
32. Donner C, Jensen HW (2006) A spectral BSSRDF for shading human skin. Rendering techniques 2006, pp 409–418
33. Donner C, et al (2008) A layered, heterogeneous reflectance model for acquiring and rendering human skin. ACM Trans Graph (TOG), vol. 27. No. 5. ACM
34. Ma W-C, et al. (2007) Rapid acquisition of specular and diffuse normal maps from polarized spherical gradient illumination. In: Proceedings of the 18th Eurographics conference on rendering techniques. Eurographics Association
35. Ragheb H, Hancock ER (2008) A light scattering model for layered dielectrics with rough surface boundaries. Int J Comput Vis 79(2):179–207
36. Dahlan HA, Hancock ER (2016) Absorptive scattering model for rough laminar surfaces. In: 2016 23rd international conference on pattern recognition (ICPR). IEEE
37. Oren M, Nayar SK (1994) Generalization of Lambert's reflectance model. In: Proceedings of the 21st annual conference on computer graphics and interactive techniques
38. Jensen HW, et al (2001) A practical model for subsurface light transport. In: Proceedings of the 28th annual conference on computer graphics and interactive techniques. ACM
39. Dutta A (2010) Face shape and reflectance acquisition using a multispectral light stage. University of York, Diss
40. Fitzpatrick TB (1988) The validity and practicality of sun-reactive skin types I through VI. Archives of dermatology, vol. 124. No. 6. American Medical Association, pp 869–871
41. Dahlan HA, Hancock ER, Smith WAP (2016) Reflectance-aware optical flow. In: 2016 23rd international conference on pattern recognition (ICPR). IEEE
42. Shirakabe Y, Suzuki Y, Lam SM (2003) A new paradigm for the aging Asian face. Aesthetic Plast Surg 27(5):397–402
43. Liew S, et al (2016) Consensus on changing trends, attitudes, and concepts of Asian beauty. Aesthetic Plast Surg 40(2):193–201
44. Vashi NA, De Castro Maymone MB, Kundu RV (2016) Aging differences in ethnic skin. J Clin Aesthetic Dermatol 9(1):31
45. Talakoub L, Wesley NO (2009) Differences in perceptions of beauty and cosmetic procedures performed in ethnic patients. Seminars in cutaneous medicine and surgery, vol 28. No 2. Frontline Medical Communications
46. Rawlings AV (2006) Ethnic skin types: are there differences in skin structure and function? Int J Cosmet Sci 28(2):79–93
47. Fink B et al (2012) Colour homogeneity and visual perception of age, health and attractiveness of male facial skin. J Eur Acad Dermatol Venereol 26(12):1486–1492
48. Matts PJ, et al (2007) Color homogeneity and visual perception of age, health, and attractiveness of female facial skin. J Am Acad Dermatol 57(6):977–984
49. Vapnik V (2013) The nature of statistical learning theory. Springer Science & Business Media, Berlin
50. Calin MA, Parasca SV (2010) In vivo study of age-related changes in the optical properties of the skin. Lasers Med Sci 25(2):269–274

Index

© Springer Nature Switzerland AG 2020
J.-D. Durou et al. (eds.), *Advances in Photometric 3D-Reconstruction*,
Advances in Computer Vision and Pattern Recognition,
https://doi.org/10.1007/978-3-030-51866-0

Printed in the United States
by Baker & Taylor Publisher Services